P9-DZZ-918

OUR ONLY HOME

Also by His Holiness the Dalai Lama

A Call for Revolution: A Vision for the Future

The Book of Joy: Lasting Happiness in a Changing World

An Appeal to the World: The Way to Peace in a Time of Division

My Spiritual Journey

HIS HOLINESS THE
DALAI LAMA
and
FRANZ ALT

OUR ONLY HOME

A CLIMATE APPEAL TO THE WORLD

HANOVER
SQUARE
PRESS

Recycling programs for this product may not exist in your area.

ISBN-13: 978-1-335-21846-9

Our Only Home: A Climate Appeal to the World

First published in 2020 in Germany by Benevento Publishing, a brand of Red Bull Media House GmbH.

English translation by Peter Reif.

This edition published by arrangement with Harlequin Books S.A.

Library of Congress Cataloging-in-Publication Data has been applied for.

Hanover Square Press
22 Adelaide St. West, 40th Floor
Toronto, Ontario M5H 4E3, Canada
HanoverSqPress.com
BookClubbish.com

Printed in U.S.A.

Table of Contents

I. Introduction by Franz Alt 13

1. Life is holy 15
2. Climate emergency 18
3. The world's most likeable person 19
4. Mankind's survival is at stake 21

II. The Third World War against Nature 23

1. Mankind is losing control 25
2. Four hundred million climate refugees 31
3. Millions take to the streets with Greta 32
4. What could rescue look like? 34
5. Dare for a future 36

III. Save the Environment—The Dalai Lama's Climate Appeal to the World 39

1. Buddha would be Green—me too, I am Green 41
2. Enviromental education 42

3. Universal responsibility 43
4. The revolution of compassion 44

IV. Franz Alt's Interview with His Holiness the Dalai Lama 47

1. The purpose of life is to be happy 49
2. We are all children of one world. 52
3. Without humans the earth would be doing better 55
4. Himalaya's glaciers are vanishing 59
5. A nuclear war would be the last in the history of mankind 65
6. More education of the heart 69

V. The Solar Age Begins—The Sun Wins 73

1. Solar energy is social energy 75
2. We should lock up the politicians 78
3. Rebirth calls for environmental protection 80
4. Buddha: "We are what we think". 82
5. Greta: "Our house is on fire" 85

VI. The Mountains Here Are as Bald as a Monk's Head .91

1. Plant trees 93
2. Ethics is more important than religion 102
3. Vegetarianism helps the climate 104

4. Buddhists disapprove of killing as a sport 106

VII. The Sheltering Tree of Interdependence— A Buddhist Monk's Reflections on Ecological Responsibility. 113

VIII. For a Solar Age—Epilogue by Franz Alt. . 127

1. Reconciling economy and ecology 129
2. There is no matter 132
3. In depth all life is one 135
4. No child should starve to death. 136
5. Disarming instead of rearming 140
6. Economizing with nature, not against it 143
7. Deeds are evidence of the truth 151

IX. Ten Commandments for the Climate . . . 155

X. What Can I Do? 163

1. Choose wisely 165
2. Is there still hope for us? 169

Acknowledgments 175

I.

Introduction by Franz Alt

1. Life is holy

"I am also an ardent supporter of environmental protection. We humans are the only species with the power to destroy the earth as we know it. Yet, if we have the capacity to destroy the earth, so, too, do we have the capacity to protect it.

"It is encouraging to see how you have opened the eyes of the world to the urgency to protect our planet, our only home. At the same time, you have inspired so many young brothers and sisters to join this movement."

This is what the Dalai Lama wrote to teenage

Swedish climate activist Greta Thunberg on May 31, 2019.

In the meantime, Thunberg has been received by the Pope and by former US President Barack Obama. She spoke before the United Nations, before the French Parliament, at two world climate summits, as well as at the World Economic Forum in Davos. She was awarded the Alternative Nobel Prize, was invited by the US Senate to speak and honored with Amnesty International's Ambassador of Conscience Award. But what has really changed? When young people were taking to the streets every Friday, where were the adults?

Barack Obama said to the shy, calm and serious young lady: "You and me, we're a team." Her simple answer: "Yes." Her motto seems to be: be humble. Before the UN summit, however, amidst tears and in a trembling voice, she flung her anger at politicians across the world:

"You have stolen my childhood. You are failing us. People are suffering, people are dying, entire ecosystems are collapsing." With clenched fist she went on: "We are in the beginning of a mass extinction, and all you can talk about is money and fairy tales of eternal economic growth. How dare you continue to look away and come here and say you're doing enough when the politics and solutions needed are still nowhere in sight?" Greta's curse! For a moment politicians had become students to whom the riot act was read, and the student had become a teacher.

In Germany, however, politicians straightaway suggested demonstrating on Saturdays, respectfully leaving the matter up to the old pros. This complacent reaction showed how powerful courageously spoken truth can be. The young protesters see through politicians who sell their responsibility for profits of large companies.

Up to now, outrage expressed by adults at the destruction of our planet is far from being loud enough. Global warming is a worldwide catastrophe, unprecedented in human history, because we are pursuing growth for the sake of growth. So we are growing poor. Prosperity gains are decreasing, while economic growth is still increasing. We have forgotten to ask: Growth, for what and for whom? We have been blind to the ecological consequences of this drive.

Greta Thunberg and her followers aim to wake us up. Maybe just in time.

A tsunami is rolling toward us. But many of us are still closing our eyes, plugging our ears and covering our mouths in the face of the danger, like the three famous Japanese monkeys.

After Greta's speech before the UN, the magazine *Der Spiegel* wondered: "Could this be the only sensible person in a crazy world?" They suggested that her words may one day be considered "a key speech of the early twenty-first century."

2. Climate emergency

In developing countries, in the first half of 2019 alone, millions have lost their homes as well as their belongings due to the effects of global warming. The poorest in particular. Global warming has already reached them; they have to fight for survival in the greenhouse. How can a religious leader and spiritual teacher help in such a situation?

Over the past thirty-eight years, I have been able to meet the Dalai Lama forty times and conduct fifteen television interviews with him on the subjects of peace, human rights, environmental and climate protection. This book appeals to the world to support the young climate activists to take a more active role in protecting this planet and to politicians to urgently tackle the global warming caused by climate change.

Besides Greta, wise and courageous women like Kenyan scientist and Nobel Peace Prize laureate Wangari Maathai or Indian agronomist and Alternative Nobel Prize winner Vandana Shiva motivate us to implement ecological transition and develop a sustainable ecological market economy, according to the motto, "eco-social instead of radically free-market."

In our conversation, it was against the spiritual background of the current problems that the Dalai Lama emphasized that we must re-examine ethically what we have inherited, what we are responsible for and what we will pass on to coming generations.

3. The world's most likeable person

Seldom in fifty years of journalistic experience have I spoken with such an empathetic, likeable and humorous person. No one has laughed more than he did. It is no coincidence that surveys show he is considered to be the happiest person worldwide. To this religious leader, interreligious ethics have become increasingly important in recent years. And what he says today distinguishes him from other religious leaders: "Ethics is more important than religion. We are not born a member of a particular religion. But ethics is innate in all of us." In his lectures worldwide, he speaks more and more frequently about "secular ethics beyond all religions." Albert Schweitzer called the same concern "Reverence for All Life." In this book, the Dalai Lama speaks of ecological ethics.

This code of secular ethics breaks down national, religious and cultural boundaries and outlines values that are innate in all people and generally binding. These are not external, material values, but inner values such as mindfulness, compassion for all creatures, mental training, as well as the pursuit of happiness. "There is no justice without compassion and mercy. If we want to be happy ourselves, we should practice compassion, and if we want others to be happy, we should also practice compassion. We all prefer to see smiling rather than glum faces," the Dalai Lama says.

One of the central beliefs of the Dalai Lama: in our pursuit of happiness and our desire to avoid suffering, we are all born equal and desire a happy and meaningful life. This results in mankind's greatest achievements. Therefore, we should think and act on the basis of deeper human values based on a sense of the oneness of humanity, with the aim of creating a more compassionate society.

In our conversations, for the individual, human dignity is the highest value and public welfare the highest collective value, since life is holy.

4. Mankind's survival is at stake

In times of global warming, extinction of species and increasing water emergency on our planet, the values of international cooperation and "universal responsibility," which the Dalai Lama is calling for ever more urgently, are particularly important. In this book, as never before, he urges politicians to urgently act after more than twenty international climate conferences, since nothing less than the survival of our planet and the sanctity of life are endangered. Even today there are regions on our planet that are scarcely reminiscent of our good old world.

It has been the vision of the Dalai Lama to make

his home country, Tibet, into the world's largest nature preserve, in accordance with the ancient Tibetan Buddhist tradition: "Tibet must and can become a demilitarized sanctuary of peace and nature."

Technology alone will not save us. Only if we combine technology with ethical responsibility can we—perhaps!—still prevent the worst of global warming. Yesterday, our beautiful blue planet was still a natural paradise in many areas. Today it is already withered in many places and tomorrow it will be uninhabitable in many regions, if we just carry on as before. But there are always alternatives. All the problems created by people can also be solved by people.

In another "Appeal to the World" in the middle of the Coronavirus crisis, the Dalai Lama pointed out that despite Coronavirus, we must not forget global warming and climate change. Because despite the pandemic, climate change is not taking a break.

Now we have the opportunity to organize a green sustainable economy worldwide while rebuilding the economy. Scientists are feverishly looking for a vaccine against the Coronavirus. We already have the vaccine against the fossil pandemic and climate change: the rapid switch to 100% renewable energies.

II.

The Third World War against Nature

1. Mankind is losing control

What does our home planet look like in the year 2020?

Rain forests everywhere are on fire, deserts are spreading on all continents, icebergs are melting, global warming means millions of climate refugees: Are we beyond hope?

In 2019, after four years of dry weather, Australia experienced the worst drought in its history. Some villages in New South Wales now have to be supplied with drinking water. Seawater desalination plants are already supplying a quarter of Sydney with water, while India is groaning under temperatures of more than 122 degrees Fahrenheit. In Europe thousands of elderly people are dying of the heat wave, already the second after the very hot summer of 2018. In Brazil, in August 2019, twice as much forest was burning as the year before, and the deforestation rate is 222 percent higher than in August 2018. Brazilian President Jair Messias Bolsonaro calls Catholic bishops and priests who oppose these arsonists "a rotten part of the Catholic Church."

Autumn 2018 was the most expensive disaster of that year worldwide and the worst fire in US history. It is both sad and ironic that the city of Paradise in California has burnt down: eighty-five people lost their lives, more than 18,000 houses and buildings on 62,000 hectares were destroyed. What a symbol: "Paradise" became hell on earth, a ghost town and a flaming hell. Three hundred thousand people had to be evacuated. The damage amounts to about fourteen billion dollars.

We experience devastating fires, heat-related deaths, drowning climate refugees, dying seas, rising sea levels, climate conflicts, polluted air, prohibitions on driving, economic collapse, water catastrophes, epidemic alarms. Global warming is far worse than we are willing to admit. The slow pace of climate change is yesterday's fairy tale. No place on earth will remain untouched and no life unchanged. We are experiencing the greatest mass extinction of the past sixty-five million years.

When I was born in 1938, the climate system still seemed intact; today it is completely out of control—within a lifetime. Caused by us humans, who are behaving like pyromaniacs. In the period between a baptism and a funeral, we have driven our planet to the edge of the abyss. The one million refugees who came to Germany in 2015 were also displaced because in their homeland, climate change and drought were among the catalysts of the civil war.

The environment is on its last legs, but in Germany, car manufacturers report record sales of large SUVs. Germans love both: the forest and the car—really an ambivalent attitude. Not just Germany, half the world is caught in the car trap.

Since 1945, 120 million people worldwide have been killed by cars. That is twice the deaths in the Second World War. And there should be no alternative? I haven't had a car for ten years now. I cover 98 percent of my journeys by public transport, where

I'm 100 times safer than in a car. Besides, I find time on the train to write my books and articles. And I also take care of the environment. Today a car consumes resources that are lost forever, and it costs an awful lot of money, yet stands around 90 percent of the time, rusting away. Our automobiles are in fact immobile. All that is the opposite of smart mobility. Without ecologically conscious traffic, there will be no turnaround in energy policy. I can easily imagine that we'll share cars and drive electrically in the future.

"Mankind is losing control of the state of the earth," warns Stefan Rahmstorf, professor and co-chair of the research department Earth System Analysis at the Potsdam Institute for Climate Impact Research and advisor of the chancellor. Climate research scientists have been mistaken in only one point during the past decades: the climatic disaster is approaching much faster than they predicted.

Glaciologists admit that the ice today is melting three times as fast as they had feared only ten years ago. Which means that this century the sea level is rising not by a few centimeters but by some meters.

So in any case there will be even more sea. And that means not only half of Bangladesh will be uninhabitable, but also New York and Shanghai, Hamburg and Bremen, Mumbai and Calcutta, Alexandria and Rio are affected. Every fourth African lives on the coast and will lose the ground below his feet if we do not stop global warming. The latest report of the Intergovernmental Panel on Climate Change predicts that today's flood damage will increase by 100 to 1,000 times by the end of this century. The rising sea level is the sword of Damocles of global warming.

Worldwide, coral reefs are dying faster than predicted. The rate at which species are becoming extinct is breathtaking. Every day we are losing 150 animal and plant species, says renowned US biology professor E. O. Wilson. We are the first generation to mess with God's creation. We play evolution backward.

Glaciologists expect a rise of the sea level by up to seventy meters, should the entire Greenland ice melt.

The target of the Paris Climate Summit to stay at 34.7 degrees Fahrenheit of warming will never be met if we continue at the current rate. Right now we are on pace for up to forty-one degrees, which means forty-six to forty-eight degrees on land, with resultant African, not South European, climate in Europe. There are still politicians and journalists who want to deny the facts provided by climate research or dismiss them as "alarmist." But this will not stop global warming. It is not a matter of faith, but simply of physics and science.

The fact is that we are waging a Third World War against ourselves since we are part of nature. Again Professor Rahmstorf: "While climate researchers have been projecting global warming quite correctly for half a century, they were wrong concerning pace and magnitude of some trends. However, they did not overestimate them, but underestimate."

Even prognoses that up to now have had little reliability like the loss of the rain forest or melting permafrost turn out to be disastrous realities. The climate tipping points of the earth system are becoming more likely. Afterward we would have no chance of survival on this planet as Homo sapiens. Stephen Hawking would be right in his prognosis: in 100 years man will have disappeared from this planet. Professor Rahmstorf explains the tipping points of the climate disaster in a quite simple and popular way, perfectly understandable, like this: "Imagine pushing a cup full of coffee over the edge of your desk. On the first centimeters little happens, the cup just changes its position. But sometime it will tilt, fall down and pour its contents onto the carpet." There is no way back for the coffee.

Eckart von Hirschhausen, medical practitioner and comedian: "The earth has a serious infection with a herd of the species Homo sapiens."

2. Four hundred million climate refugees

According to the World Bank, about 100 to 140 million climate refugees will wander around our planet by the year 2030, looking for the nearest watering hole. And the UN expects more than 400 million climate refugees by 2050, if we do not stop global warming. The climate crisis will thus become a democracy crisis and a social crisis, even the threat of war. That is at stake: Wars for oil or peace through the sun? Nobody will ever wage war for the sun. The temperature there is around 194 million degrees Fahrenheit.

What do we have to learn to finally overcome the climate crisis? There is not much time left.

Our biggest problem is not the enormous number of people, but a lack of humanity.

3. Millions take to the streets with Greta

What a crazy world we live in! A seventeen-year-old girl holds the climate mirror up to us. And what do we see? Ourselves! The Dalai Lama says: "The belief in rebirth calls for more environmental and climate protection. Because we will come back to this planet and, also for that reason, want a good climate and a healthy earth." What a reminder! On March 15, 2019, Greta, in her simple but determined language ("Our house is on fire," "I want you to panic") drove 1.6 million young people onto the streets and initiated the first worldwide climate strike on September 20, 2019, with over six million protesters. Greta is like the child in Andersen's fairy tale "The Emperor's New Clothes," calling to the adults: "But he has nothing on at all." The seventeen-year-old girl is calling: "No life without a good climate." The world rubs its eyes, begins to wake up and realizes: the child is right. Curiously, these days children and young people who demonstrate for a good climate behave more like grown-ups than the adults.

When the German speaker of the "Fridays for Future" movement, geography student Luisa Neubauer, spoke to the shareholders of the energy supplier RWE, her microphone was taken away for a moment. What a fear of the truth.

A week later, millions of mainly young people took to the streets, in Wellington and Vienna, in Stockholm and New Delhi, in order to demonstrate—in the spirit of Greta—for better climate protection. In Italy alone they were over a million. In Naples, placards read, "We want hot pizza, but not a hot planet."

In Germany, after these mass-participant demonstrations, the Greta effect also had effects on political parties: surveys showed that with 27 percent, the Greens were neck and neck with the Christian Democrats (CDU) and Bavarian Christian Social Democrats (CSU), who were once advocates of nuclear power. SPD, the party advocating coal, just reached 13 percent.

4. What could rescue look like?

First: in 2035 at the latest, we will reach 100 percent renewable energy. This is no problem, because the sun sends us 15,000 times more energy than we all are at present consuming. This is environmentally friendly, forever and free. The solution is in the sky: think wind energy, hydropower, biomass and geothermal energy. Sun and wind don't send out any bills. What an incredible economic advantage of future ecological energy supply. This insight is as simple and irrefutable as if it came from Greta. But why do 90 percent of all roofs in Germany still stand around uselessly without solar panels?

We all have learned at school: everything revolves around the sun. But why don't we see the light? We must finally open up for energy from very far above, for energy from the boss himself. Without the sun, there's no life.

Second: rapid withdrawal from coal and a price on CO_2. This should be possible before 2038, Greta thinks. Transition to renewable energy creates far more jobs than are lost by the abandonment of coal. Meanwhile French President Emmanuel Macron also agrees with Greta. He, too, says: "Our house is on fire." When do we finally call the fire brigade? But instead of extinguishing the flames, we are discussing the price of the firefighters' water.

Third: rapid entry into electric mobility (whether e-bikes or electric cars) and a doubling of public transport. China, Norway, California and the Netherlands are showing the way.

Fourth: switching to organic farming. More and more municipalities renounce the use of pesticides and glyphosate. Our consumer behavior is politics with a shopping cart. What is at stake, finally, is elementary: fertile soil, drinking water, clean air, a moderate climate and forests that are good for our souls, which are under stress, and store the excessive CO_2.

Fifth: worldwide reforestation and greening of the deserts. A study carried out by a Zürich university, ETH, has recently shown that reforestation in the USA, in Russia, China, Brazil and Canada could make up for more than two thirds of man-made CO_2 emissions. What are we waiting for? The child and youth organization "Plant-for-the-Planet," which was initiated in Starnberg by eight-year-old Felix

Finkbeiner, has shown the way. In the past fifteen years, over thirteen billion trees have been planted worldwide. Children and young people are already playing a major role, when a good future is at stake. We adults must finally learn to take the justifiable fears of our children and grandchildren seriously.

5. Dare for a future

Politicians and parties still have time to campaign for this survival program. Motto: the climate catastrophe endangers our well-being and our lives. Smart climate politics secures and preserves our well-being and guarantees the future of our children and grandchildren. We must dare for a future. People's courage has always been decisive. For example, in the French Revolution in 1789, the suffragette movement in the early part of the twentieth century or the peaceful revolution in Germany in 1989. Often there are only a few at the beginning who dare a lot. And like Greta, they were all alone.

We still have the choice. But time is running out. The energy transition is expensive, that is true. But no energy transition will cost mankind's future, says

German politician Wolfgang Schäuble. It is not a question of a sacrifice or renunciation, but rather a question of desire for the future on this unique and wonderful earth.

In this book, the Dalai Lama shows how alternative forms of economic activity can be developed. A Buddhist economy based on the common good can be very helpful due to the principles of mindfulness, nonviolence, compassion and modesty—for our personal as well as our political and economic behavior. Is this book reason for alarmism? No way! It is a declaration of love to the future.

Franz Alt
Baden-Baden, Autumn 2019

III.

Save the Environment—
The Dalai Lama's Climate Appeal
to the World

1. Buddha would be Green—me too, I am Green

Buddha was born as his mother leaned against a tree for support. He attained enlightenment seated beneath a tree, and passed away as trees stood witness overhead. Therefore, were Buddha to return to our world, he would certainly be connected to the campaign to protect the environment.

Speaking for myself, I have no hesitation in supporting initiatives that are related to protecting the environment because threats to our environment are a question of our survival. This beautiful blue planet is our only home. It provides a habitat for unique and diverse communities. Taking care of our planet is to look after our own home.

We can no longer keep exploiting the resources of this earth—the trees, the water and the minerals—without any care for coming generations. It is common sense that we cannot survive if we keep working against nature. We must learn to live in harmony with nature.

If we compare damage to the environment to war and violence, it's clear that violence has an immediate impact on us. The trouble is that damage to the environment takes place more stealthily, so we don't see it often until it is too late. We have reached a tipping point in global warming.

2. Environmental education

Environmental education about the consequences of the destruction of our ecosystem and the dramatic decrease in biodiversity must be given top priority. But creating awareness is not sufficient; we must find ways to bring about changes in the way we live. I call on the younger generation—be rebels demanding climate protection and climate justice because it is your future that is at stake.

One of the most positive recent developments has been the growing awareness that we have to act. Seventeen-year-old Greta Thunberg, the teenage environmental activist who insists we heed scientists' warnings and take direct action, inspires me. Millions of young people have been moved by her example to protest governments' inaction over the climate crisis. She is correct to say, "No one is too small to make a difference." I wholeheartedly support "Fridays for Future," the movement she initiated.

I am encouraged to see young people's determination to bring about positive change. They are confident of making a difference, because their efforts are based on evidence and reason.

More and more people understand that the survival of humanity is at stake. Simply meditating or praying for change is not enough. There has to be action.

3. Universal responsibility

We seven billion human beings must learn to live together. It is no longer enough to think only of "my country," "my people," "us" and "them." We must each learn to work for the benefit of all human beings.

We are social animals born with a sense of belonging to a community. We have to realize that just as our future depends on others, theirs depends on us. Our world is deeply interdependent, not only in terms of our economies but also in facing the challenge of climate change.

We have to appreciate that local problems have global ramifications from the moment they begin. The climate crisis affects the whole of humanity.

Island states like Fiji, the Marshall Islands, the Maldives and the Bahamas have shown that collectively we can make a difference. The 2015 Paris Agreement to combat climate change signed by 196 countries was a good start that has to be followed through with action.

We need a sense of universal responsibility as our central motivation to rebalance our relations with the environment and with our neighbors. Appreciating the oneness of humanity in the face of the challenge of global warming is the real key to our survival.

4. The revolution of compassion

I am now eighty-four years old and have lived through many of the upheavals of the twentieth century—the destruction and suffering brought by war, but also unprecedented damage to the natural environment. Today's younger generation has the ability and opportunity to create a more compassionate world. I urge them to make this twenty-first century an era of change rooted in dialogue and a century of compassion for all the inhabitants of this planet.

Over-exploitation of our natural resources results from ignorance and greed, and a lack of respect for life on earth. Saving the world from the climate crisis is our common responsibility. We must find ways to exercise freedom with responsibility.

We need a revolution of compassion based on warmheartedness that will contribute to a more compassionate world with a sense of oneness of humanity. The entire human family must unite and cooperate to protect our common home. I hope that efforts to achieve a more sustainable way of life will meet with success.

Dalai Lama
Dharamsala, India, December 10, 2019

IV.

Franz Alt's Interview with His Holiness the Dalai Lama

1. The purpose of life is to be happy

Franz Alt: *Your Holiness, dear friend. Fifteen years ago you said to me in an interview: "The twenty-first century could become the happiest and most peaceful one in human history. I hope so for the youth." Do you still cherish that hope?*

Dalai Lama: I am hopeful that the twenty-first century could become the most important century in human history. The twentieth century experienced immense destructions, human sufferings and unprecedented environmental damages. The challenge before us, therefore, is to make the twenty-first century a century of dialogue and promotion of the sense of oneness of humanity.

As a Buddhist monk, I appeal to all human beings to practice compassion—the source of happiness. Our survival depends on hope. Hope means something good. I believe the purpose of life is to be happy.

The world's seven billion human beings must learn to work together. This is no longer a time to think only of "my nation" or "our continent" alone. There is a real need for a greater sense of global responsibility.

I feel optimistic about the future because humanity seems to be growing more mature; scientists are paying more attention to our inner values, training of the mind and the emotions. There is a clear desire for peace and concern for the environment.

Franz Alt: *The Paris Climate Summit at the end of 2015 was the beginning of a new reality. For the first time the world may have seen itself as a world family. There, all governments worldwide and the European Union have committed themselves in writing not to increase global warming by more than 1.5 degrees Celsius, two degrees at the most—compared with 1880 levels. But globally, we already have an increase of more than one degree. If we continue like that, global warming can reach five to six degrees, even* *in this century. I would not want to be my grandson then. Paper is patient—governments are not acting. Are you still optimistic? Can the Paris Agreement still be achieved?*

Dalai Lama: I hope and pray that the 2015 Paris Agreement will finally bring tangible results. Egotism, nationalism and violence are fundamentally wrong.

America's withdrawal from the Paris Agreement is very sad. It is important for scientists to continuously speak up about the dangers we face and alert the public. Here, the media has an important responsibility in educating the people. The gap between rich and poor is also very serious and we have to take steps to close it by helping the poor.

Any human activity should be carried out with a sense of responsibility, commitment and discipline. But if our activities are carried out with shortsightedness and for short-term gains for money or power, then they all become negative and destructive activities. Protecting our environment is not a luxury we can choose to enjoy, but it is a matter of survival.

A small temperature rise in our body creates much discomfort. Annually we have been witnessing global warming due to climate change. Recently, both America and Europe have experienced extremely hot summers and cold winters. The questions of the environment and climate change are a global issue, not just of concern to Europe, Asia, Africa or the Americas. What happens on this blue planet affects us all.

It is not sufficient to just express views and hold conferences. We must set a timetable for change.

Franz Alt: *As early as 1992 you said: "Universal responsibility is the key to human survival." What does that mean in concrete and practical terms?*

Dalai Lama: The seven billion human beings are social animals and must learn to live together. This is no longer a time to think only of "my country," "my people," "us" and "them." We live in a globalized world. Countries think about their own national interests rather than global interests, and that needs to change because the environment is a global issue. In order to protect global environmental issues, some sacrifice of national interests is needed.

2. We are all children of one world

Franz Alt: *Nationalism has been shaping our history for centuries. Is there really a possibility to overcome nationalist thinking?*

Dalai Lama: Wherever I go I emphasize that all seven billion human beings are physically, mentally and emotionally the same. Everybody wants to live a happy life free from problems. Even insects, birds and animals want to be happy.

In order to ensure a more peaceful world and a healthier environment, we sometimes point a finger at others, saying they should do this or that. But change must start with us as individuals. If one individual becomes more compassionate, it will influence others, and so we will change the world. Scientists say our basic nature is compassionate. This is very positive.

In the face of such global problems as the greenhouse effect and depletion of the ozone layer caused by chlorofluorocarbon gases, individual organizations and nations are helpless. When I was awarded the Nobel Peace Prize in Oslo in 1989, I called upon the world to assume universal responsibility. We have to learn that we are all brothers and sisters and live on one earth and under the same sun.

Unless we all work together, no solution can be found. Therefore, our key responsibility is to commit ourselves to the ethical principles of universal responsibility beyond profit and religion, and to place the well-being of all sentient beings and future generations above our egoism. Climate change is an issue that affects the whole of humanity. But if we have a genuine sense of universal responsibility as our central motivation, then our relations with the environment will be well-balanced, and so will our relations with our neighbours. Our Mother Earth is teaching us a lesson in universal responsibility.

Therefore, each of us as individuals has a responsibility to ensure that the world will be safe for future generations, for our grandchildren and great-grandchildren.

Franz Alt: *Is global warming just a political problem or can every individual do something about it?*

Dalai Lama: According to scientists, we human beings are responsible for global warming and the change in weather conditions. Logically this means that we human beings have a responsibility to solve problems that we have created.

On an individual level, we should change our lifestyles, consume less water and electricity, plant trees and reduce the use of fossil fuels, which took millions of years to form. Fossil fuels are nonreusable energy;

therefore, we must use renewable energy like solar, wind and geothermal.

As a boy studying Buddhism, I was taught the importance of a caring attitude toward the environment. Our practice of nonviolence applies not just to human beings but to all sentient beings.

What distinguishes human beings from animals? It is our specific capacity for long-term thinking. Animals only live from one day to the next, whereas our brain can think ten or even a hundred years ahead. A thousand years may be too much for us. In consequence we are equipped to make preparations for the future and plan for the long term.

3. Without humans the earth would be doing better

Franz Alt: *But is it only our shortsightedness that prevents us from treating our natural environment carefully?*

Dalai Lama: Destruction of nature and its resources results from ignorance, greed and lack of respect for the earth's living things. Today, we have access to more information, and it is essential that we re-examine ethically what we have inherited, what we are responsible for, and what we will pass on to the coming generations.

Resolving the environmental crisis is not just a question of ethics but a question of our own survival. The natural environment is very important not only for those of us alive now but also for future generations. If we exploit it in extreme ways, even though we may get money or other benefits from it now, in the long run we ourselves and future generations will suffer. When the environment changes, climatic conditions also change. When they change dramatically, the economy and many other things change as well. Even our physical health can be greatly affected.

Franz Alt: *In the past, man needed protection from his environment. Today it is the other way round. Scientists tell us: without humans the earth would be doing better.*

Dalai Lama: As someone born in Tibet, the rooftop of the world, where the world's highest peaks are to be found and Asia's great rivers originate, I have loved nature since my childhood. I have made environmental conservation one of my life's commitments and advocate protection of the environment wherever I go. Therefore, I called on all to speak out about global warming, which affects the future.

Franz Alt: *In his Encyclical on the environment, what you call "universal responsibility," Pope Francis puts like this: "The interdependence of all creatures is God-given. The sun and the moon, the cedar and the field flower, the eagle and the sparrow—the myriad of differences and inequalities is evidence that creatures are not self-sufficient, but exist only in dependence on each other and complement one another in mutual service." Does the Pope's statement match with what you think?*

Dalai Lama: I welcome Pope Francis's Encyclical on the environment. I also see similarities between the Pope's Encyclical "One human family, one common home" and my message of oneness of humanity. Since global warming and climate change affect us all, we have to develop a sense of the oneness of humanity and universal responsibility.

The metaphysics of the wise men of ancient India and the West are converging in times of ecological crisis. Technology alone will not save us. We need interdependence of ethics and technology. We need a joint plan to save the planet.

Caring for the earth is our shared responsibility. Each one of us has a moral responsibility to act, as so powerfully stated by the Pope's Encyclical.

Franz Alt: *In our previous book,* The Way to Peace in a Time of Division, *you expressed the idea that "Ethics is more important than religion." What does that mean as far as environmental policy is concerned?*

Dalai Lama: Religion should not be just limited to praying. Ethical action is more important than prayers. What is Buddha, Allah or Christ supposed to do if we human beings destroy our earth; fill the oceans with plastic so that fish, seals and whales perish; and cause rapid increase of desertification and greenhouse gases to be released into the atmosphere? Christ, Allah or Buddha is not responsible for the climate change and the destruction of the environment; it is a man-made problem. Therefore, we must take the responsibility and find solutions to the problems. That is why we need environmental ethics that focus on action and compassion for all sentient beings.

Scientists have concluded that basic human nature is compassionate. Those who grow up in a more compassionate atmosphere tend to be happier and more successful. On the other hand, scientists suggest that living with constant anger or fear undermines our immune system. Hence, compassion and warmheartedness are not only important at the beginning of our lives but also in the middle and at the end. Their necessity and values are not limited to any specific time, place, society or culture.

4. Himalaya's glaciers are vanishing

Franz Alt: *Two thirds of the Himalayan glaciers are in danger of disappearing by 2050 due to global warming. That would affect the water supply of billions of people in India and China.*

The Hebrew Bible states: "In compassion, righteousness and peace will kiss each other." And in the New Testament, Jesus says, "Be compassionate even as your heavenly Father is compassionate." Our actions are only compassionate when they result from our solidarity. In North Africa I have seen areas that were a paradise for humans yester-

day; today they are already drought-stricken, and tomorrow they will be uninhabitable. As a television journalist I've been able, for fifty years now, to observe similar catastrophic developments in India and Bangladesh. Can ice melting and global warming be stopped at all?

Dalai Lama: Billions of dollars are spent on weapons of mass destruction. If half of these billions of dollars was used to develop new technologies and wider use of renewable energy, the positive impact it would have in our efforts to limit global warming would be tremendous!

Resting hopes on the younger generation is not sufficient. Politicians, too, must urgently act. It is not sufficient to hold meetings and conferences. We must set a timetable for change. Only if they start to act now will we have reason to hope. We must not sacrifice our civilization for the greed of the few. Journalists have an equally important role. I tell them that in this modern time they have a special responsibility to bring awareness to the people—not just report on bad news, but bring people hope.

Recent studies suggest that the world is getting close to exceeding its carbon budget. Therefore, this budget must become the most important currency of our time. Politicians are gradually running out of excuses, but we must use our time wisely.

Hundreds of thousands of young people are taking to the streets in the new global youth environmental movement "Fridays for Future" to persuade politicians for better climate protection. I am encouraged to see young people's determination to create greater awareness to bring about positive change. They will succeed because their efforts are based on scientific truth and reason.

This little book is a call for action! An appeal to all politicians, opinion formers, journalists, religious leaders and to all people. Since the future of all coming generations rests on our shoulders, we must be determined to take action before it is too late.

Franz Alt: *What you have just said about politics and politicians also applies to us journalists. We must finally begin to describe the crisis as a crisis. We are facing unimaginable suffering of billions of people. Why is the climate issue a matter of survival for all life?*

Dalai Lama: I often joke that the moon and stars look beautiful, but if any of us tried to live on them we would be miserable. This blue planet of ours is a beautiful habitat. Its life is our life; its future our future. Indeed, the earth acts like a mother to us all. Like children, we are dependent on her. Our world is deeply interdependent, both in terms of our economies and the problems like climate change that challenge us all.

Even a small temperature rise in our body creates much discomfort. Scientists say even a small rise in earth's temperature is a risk to human beings, animals, agriculture, water and melting of glaciers especially in the Arctic and Antarctic, in Greenland and Alaska, in the Himalayas and in the Alps. If the world fails to halt global warming, small island states may disappear forever due to rising sea levels. Unfortunately, the poor are hit hardest in weather-related disasters.

When we see photographs of the earth from space, we see no boundaries between us, just this beautiful blue planet. This is no longer a time to think only of "my nation" or "our continent" alone. There is a real need for a greater sense of global responsibility based on the oneness of humanity.

Franz Alt: *Why do you say that Tibet is the big epicenter of climate change?*

Dalai Lama: A Chinese ecologist has described the Tibetan Plateau as the Third Pole because it is the third largest area of frozen water on the planet after the North Pole and South Pole. The effects of global warming on the Tibetan Plateau have a significant impact on the lives of more than 1.5 billion people living in the region. The same article mentioned that the temperatures on the Tibetan Plateau have increased by 1.5 degrees—more than double the global average. The Third Pole's glaciers are melting at a rate that has almost doubled since 2005. More than 500 small glaciers have disappeared altogether, and the biggest ones are shrinking rapidly, according to the research.

The Tibetan Plateau happens to be the largest water tank in the world. All the ten major rivers of Asia including the Ganges, Karnali, Brahmaputra, Indus, Sutlej, Irrawaddy, Salween, Yellow River, Yangtze and Mekong originate in the Tibetan Plateau. Over 1.5 billion people live by these waters—one fifth of the world's population. Without water there's no life. If Tibet's 46,000 glaciers continue to melt, we will face unimaginable water problems and probably water will become a key cause for conflict in the future. So the ecology of Tibet is something really important.

Franz Alt: *We were both standing at the Berlin Wall in November 1989. And in East and West Berlin the* "Mauerspechte"*—"wall-peckers"—were already hammering at this inhuman monster. People on either side of the wall gave you a burning candle and heaved you onto the remains of the wall. There you stood using these powerful words: "As sure as this wall will fall, my home, Tibet, will one day have freedom." For me, an unforgettable moment. After all, only a few months before, Chinese rulers had brutally ended the student revolt on Tiananmen Square. Afterward, we discussed your optimistic perspective in front of several thousand students at the Freie Universität Berlin. Would you repeat that sentence today while the repression in Tibet under Chinese occupation has increased since 1989? Do you remain such an optimist?*

Dalai Lama: When we came into exile, preserving our identity, language and culture was our priority. These days, when Tibetans in Tibet reveal their passion likewise to preserve their culture, Chinese hard-liners oppose it as indicative of their "separationist" motives. Despite restrictions, Tibetans' spirit remains resolute.

Things are changing and a totalitarian system has no future.

5. A nuclear war would be the last in the history of mankind

Franz Alt: *You fear wars over water between India and China. Both countries have nuclear bombs. Could there ever be a nuclear war between India and China on the water issue?*

Dalai Lama: A nuclear war would probably be the last one in human history, because nobody would be left to wage another war.

Franz Alt: *Why is it important to maintain harmony between the natural environment and the sentient beings?*

Dalai Lama: There is a very close interdependence between the natural environment and the sentient beings living in it. Thus, we share a sense of universal responsibility for both mankind and nature.

When the environment becomes damaged and polluted, there are many negative consequences. Oceans and lakes lose their cool and soothing qualities, so the creatures depending on them are disturbed. The decline of vegetation and forest cover causes the earth's bounty to decline. Rain no longer falls when required, the soil dries and erodes, forest fires rage and unprecedented storms arise. We all suffer the consequences.

Prior to the Chinese occupation, Tibet was a fresh, beautiful, unspoiled wilderness sanctuary in a unique natural environment. Sadly, in the last six decades, Tibetan wildlife and its fragile ecology have been almost destroyed by the Chinese occupation. What little is left must be protected. Every effort must be made to restore the Tibetan environment to its balanced state.

In spite of all the suffering that China has inflicted on Tibetans for over six decades, I remain convinced that most human conflicts can be solved through sincere dialogue held in a spirit of openness and reconciliation. We have learned that even enemies can become friends. I am a strong believer in nonviolence.

Franz Alt: *"The environmental catastrophes are the reflection of our combative and destructive ways of think-*

ing based on a selfish desire for prosperity and profit," you have said. Isn't a certain egoism and pursuit of wealth part of man's essence?

Dalai Lama: *[After a moment's reflection and a smile, the Dalai Lama answers.]* Material values do matter. But deeper inner values are more important than material values. In the last century, we made great material progress. But it is precisely this material progress that is now leading to environmental destruction. We need a new balance between economy and ecology, otherwise we will destroy the basis of our lives. Material progress alone cannot reduce our psychic stress, anxiety, anger and frustration.

My friend Mikhail Gorbachev is still committed to environmental issues in the international organization "Green Cross," which he co-founded. Ecology must become the smarter economy. Only then will we be able to live in a sustainable way.

As to whether the world is getting better or worse, there is growing opposition to the existence of nuclear weapons. Where no one used to talk about the environment, today it is on everyone's lips. Scientists who once only paid attention to material things are now paying attention to training the mind. I am optimistic that people are generally becoming more mature.

I already described us as selfish, that's right. But we should be wise selfish rather than foolish selfish. Think less of "I" and think more about others' well-being. You get maximum benefit. So that is wise selfishness.

Franz Alt: *You talk about prioritizing for environmental education—from kindergarten to high schools, but also in the universities. Why is it important to start so early in children's education?*

Dalai Lama: Every child should learn in school that their own future and happiness will always depend on the future and happiness of others. Even in kindergarten, children can learn that all seven billion human beings have the right to be happy. We all live on the same planet, under the same sun, and breathe the same air. Today, the world needs environmental ethics education based on a deeper understanding that transcends religion. At school, children can learn to appreciate it.

Environmental education must be given top priority as we all have become witnesses to the destruction of our ecosystem and a dramatic decrease in biodiversity. Environmental education means learning to maintain a balanced way of life. In order to maintain a universal appeal, such ethics need to have a secular basis.

When I came from Tibet to India in 1959, I had no idea of the problems with the environment. When I first heard, "You cannot drink this water," I was surprised that it was polluted. In Tibet, passing through waters, by a stream, we always would enjoy it. No problem. I learned about pollution and gradually about ecology. I now feel a deep concern about the environment, as it has become a question of our survival. I learned about this through awareness—not through meditation, but through awareness with help from experts.

We may talk about going to the moon or to Mars, but we can't settle there. This earth is the only place where we can live.

6. More education of the heart

Franz Alt: *What do you mean by the phrase "education of the heart"?*

Dalai Lama: My wish is that more attention is paid to educating the heart—teaching love, kindness, peace, compassion, forgiveness, mindfulness, self-discipline, generosity and tolerance. This education is necessary from kindergarten to secondary schools and universities. I mean social, emotional and ethical learning. We need a worldwide initiative for educating the heart and training the mind in this modern age.

At present our modern educational systems are oriented mainly toward material development. Modern education is not adequate and it pays little attention to inner values. Yet our basic human nature is compassionate. Therefore, we need to develop a curriculum based on compassion and warmheartedness in the modern education system to make it more holistic.

We are not like plants; we have emotions. We need to learn to manage our emotions and achieve inner peace. Our education should include an understanding of how to achieve peace of mind. Education should teach us how to live properly, how to balance our wish for physical comfort with mental comfort. That's what's important.

We must learn that humanity is one big family. We are all brothers and sisters: physically, mentally and emotionally we are the same. But we are still focusing far too much on our differences instead of our commonalities. After all, every one of us is born the same way and dies the same way.

V.

The Solar Age Begins—
The Sun Wins

1. Solar energy is social energy

Franz Alt: *In many of our discussions we have often talked about solar energy. The sun sends 15,000 times more energy to Earth than we currently consume. And the power of the sun is available at no cost, environmentally friendly, worldwide and forever. The solution of the energy problem is in the sky. Besides, there is also wind energy, hydropower, bioenergy, geothermal energy and ocean energy. The world is full of energy. We can make use of the entire symphony of renewable energies. In some countries, such as Costa Rica or Iceland, the entire energy industry is already renewable. Worldwide, even one third of electricity production is renewable. There is no lack of knowledge or* *technologies, but only of rapid implementation. Why does transition take so long?*

Dalai Lama: You told me that Germany today produces as much as 50 percent green electricity. In 2000 it was only 5 percent. This shows even industrialized countries with high energy consumption can switch to renewable energy. I understand that storage technologies for solar and wind power are now more developed. Besides, the sun and wind do not send an invoice. That means they are nature's gifts, which we should use much more in the future. Solar and wind energy are already the more economical sources of energy worldwide. So we don't need nuclear or coal-fired power plants. We are at the beginning of a worldwide solar revolution.

We must change our lifestyle and heavy reliance on old energy. So, there must be more government incentives to renewable energy companies and to the common people to use renewable energies.

Franz Alt: *May I kindly ask, dear friend: we can already build houses and factories that produce more green energy than they use. The sun shines on every roof. We* *know that in the future, ships and airplanes will use solar-generated hydrogen. In the past twenty years, we have created over eleven million sustainable jobs worldwide through renewable energy. The International Renewable Energy Agency estimates that the solar energy transition will create twenty-five million new jobs by 2030. But why is the energy transition so slow?*

Dalai Lama: New technology has always taken a relatively long time to make its complete breakthrough. More and more companies are producing electric cars. But if the cost of the car is too expensive, then only the very rich can buy them. So these cars must be more affordable. Similarly, other renewable energy must be made more affordable, especially to the poorer section of the community, who are most vulnerable due to climate change. I have learned from scientists that the use of solar power and wind power worldwide has increased over the years. We are making good progress. As I said earlier, there must be more education and incentives on the use of renewable energies.

2. We should lock up the politicians

Franz Alt: *We can already produce solar power in Africa or in Chile at two and a half cents per kilowatt-hour. The government of Saudi Arabia wants to produce solar power at one cent by 2025 in the largest solar power plant in the world. These are all gifts from the sky. This is also the most efficient development policy of all time. Solar energy being already social energy. Renewable energy is the decisive step to prosperity for all. With cheap energy, the economy can be developed in the poor countries of the Southern Hemisphere and the causes of flight can be eliminated. How can we get politicians to swiftly implement what they decided upon in Paris?*

Dalai Lama: *[Breaking into his world-famous laugh and grinning, the Dalai Lama replies.]* Perhaps you should lock up the world's most important politicians in a room for a while and pipe carbon dioxide into it until they realize what climate change really means. Then, they would probably very soon feel what greenhouse gases are doing to us humans *[still laughing quite loudly]*.

I very much appreciate the worldwide activities for the energy transition and for more environmental protection. Because I often have the impression that politicians do not take climate and environmental protection seriously enough. Ignorance is the number one enemy.

Scientists say that due to global warming, many parts of the world could become desert. This is very serious.

Dr. Yuan T. Lee, a Taiwanese Nobel laureate in Chemistry, told me that after another eighty years the world will be like a desert. He said that water resources are already diminishing alarmingly. Therefore, we all need to modify our lifestyle, abandoning fossil fuels and turning to renewable sources of energy.

One of my dreams, perhaps an impossible dream, is to harness the solar potential of places like the Sahara Desert and to use the power to run desalination plants. The sweet water thus produced could green the desert and produce food crops. It is a project that would have widespread benefits and would function on a scale that requires global cooperation.

So, taking care of the environment, taking the necessary steps to reduce global warming, is a serious matter. I'm a monk so I have no children, but people who have children have to think about how life will be for them and their grandchildren. We're at the start of the twenty-first century, but we should also be looking ahead to how things might be in the twenty-second and twenty-third centuries.

3. Rebirth calls for environmental protection

Franz Alt: *Could perhaps a consciousness of reincarnation also help? Whoever knows that he will come back, wants a healthy planet. We know today that Jesus spoke repeatedly of rebirth in his Aramaic mother tongue. According to the Jewish philosopher Schalom Ben-Chorin, the belief in rebirth was a popular belief. But later the Christian bishops eliminated these words of Jesus by majority vote. Rebirth is taken for granted in Asian religions. But in all religions and cultures, there are people who remember past lives. Cultures and religions that believe in rebirth should, if only for selfish reasons, stand up for environmental protection, because in their next lives they also need a hospitable earth. Does environmental and climate protection have a higher priority in Asian rebirth religions than in Western religions?*

Dalai Lama: Yes it does. But hardly in practical politics. In practical environmental behavior, very little can be felt so far. Think of China's environmental problems. Or of Japan's nuclear power plants. And coal-fired power plants in India.

It is true, however, that a person who believes in rebirth naturally wants an environmentally friendly planet in our next life. Me too *[laughing]*.

We can no longer keep exploiting the resources of this earth—the trees, the water and the minerals—without any care for the coming generations. It is common sense that we cannot survive if we keep working against nature. We must learn to live in harmony with nature. As a Buddhist monk who believes in rebirth, even for selfish reasons, we must pay more attention to our planet. Because we will come back. And all of us would like to live on a healthy earth. The belief in rebirth calls for more environmental and climate protection.

Franz Alt: *Wise men of all cultures were convinced of reincarnation. For example, the German philosopher Schopenhauer in the nineteenth century or Pythagoras in ancient Greece as early as twenty years before Buddha. The Christian church father Origen of Alexandria was also convinced of reincarnation. The Jewish religious scholar Schalom Ben-Chorin writes: "In the days of Jesus the belief in reincarnation was popular belief." Today's*

Christian Western world is the only region on our planet where reincarnation is officially denied. You, dear friend, say, "Spirituality is the essential key to our survival." Could you give reasons for this statement?

Dalai Lama: I have often said that in keeping with the tradition of Tibetan Buddhist culture, all sentient beings have been our mothers. The entire Buddhist spirituality is characterized by this realization. All sentient beings are connected by a maternal bond. This is the basic truth of awakening, enlightenment and realization. We are all interconnected in the universe, and from this, universal responsibility arises. Jesus knew this spiritual law called "karma" in Buddhism, and talked about it without using the word *karma.* It is a spiritual law that "you reap what you sow." Things entirely depend on your effort, your action. So things change through action and not by prayer. We must act to create positive karma. Positive karma means positive action.

4. Buddha: "We are what we think"

Franz Alt: *Buddha said, "We are what we think. Everything we are arises from our thoughts. Our thoughts* *shape our world." In the last few months, teenager Greta Thunberg from Sweden has proved with her firm will what a single person can achieve. She started her school strike in the summer of 2018 and on a Friday all by herself sat down in front of the Reichstag in Stockholm. On her protest poster it said, "School strike for the climate." On the following Friday, four students, girls and boys, were sitting next to her. And today, hundreds of thousands follow her in over a hundred countries: on March 15, 2019, it was 1.6 million; on September 20, 2019, over six million.*

The young lady spoke at the World Climate Summit in Poland, met the Pope, was nominated for the Nobel Peace Prize and in Sweden was named "Woman of the Year." Time Magazine *counts her among the 100 most influential people worldwide. When asked why she is fighting for the climate, she says: "I know what's at stake—the survival of mankind. And I consider it my moral duty to do everything I can to avert the worst. At first, I tried to inspire others, but nobody wanted to get involved, so I started alone. We can achieve a lot, if a lot of people join in." She also says that she took part in many demonstrations for the climate before, but nobody reported on it. Only when she had the idea of a school strike and put it into practice did it become a world issue. And today she tells the politicians: "We're going on strike until you* *act. Together we will change the world." The shy teenage girl fell ill because she could no longer stand the images of plastic mountains in the oceans. Her mother says: "After her illness, Greta sees things that others don't see: the CO_2 of airplanes and cars and coal-fired power plants. She sees that we are transforming the atmosphere into an invisible gigantic garbage dump." As a Buddhist, how do you explain today's global Greta effect that no scientist could have devised?*

Dalai Lama: I really appreciate Greta Thunberg's efforts to raise awareness of the need to take direct actions. People like her are realistic. We should encourage them.

The young Greta's motivation to create awareness on global warming among school children is a remarkable achievement. Despite being very young, her sense of universal responsibility to act is wonderful. I support her "Fridays for Future" movement.

I believe that every individual has a responsibility to help guide our global family in the right direction. Good wishes alone are not enough; we have to assume responsibility. Large human movements spring from individual human initiatives.

The youth of the twenty-first century are the

planet's real humanity now. They have the ability and opportunity to bring change, to create a century of peace, dialogue and compassion. Even as global warming increases in intensity, they can work together in the spirit of brotherhood and sisterhood to share and find solutions. They are our real hope.

Ideas may travel from the top down, but the movements that put them into effect have to work from the bottom up. Therefore, I am encouraged to see young people trying to bring about positive change. Confident because their efforts are based on truth and reason—therefore, they will succeed.

So now, the generation of the twenty-first century, the young brothers and sisters must take a more active role in protecting ecology and our home.

5. Greta: "Our house is on fire"

Franz Alt: *Is Greta, together with the thousands of pupils and teenagers that have begun to follow her, right when they call out to us old people: "We are loud because you're stealing our future"? Greta says: "Our house is on fire." Is that exaggerated?*

Dalai Lama: The young climate activist is right. Scientists and environmental campaigners have been selflessly and tirelessly putting effort into creating a better environment for the world so that future generations will be able to live a healthy, happy life. The 2015 Paris Agreement was signed by leaders of 196 countries to combat climate change and limit temperature rise to "well below" 3.6 degrees Fahrenheit, and it is a source of hope and encouragement. They are now committed to more effective climate protection. If millions of young brothers and sisters all over the world go on strike because politicians are inactive, that's a sign that something is not in order.

Climate change is not the concern of just one or two nations. It is an issue that affects all humanity. This beautiful planet is our only home. If, due to global warming or other environmental problems, the earth cannot sustain itself, there is no other planet we can move to and live on. We have to take serious action now to protect our environment and find constructive solutions to global warming.

Franz Alt: *How can we motivate politicians as well as businesspeople to do more for the environment and the climate than they have done so far?*

Dalai Lama: In recent months, millions of young brothers and sisters have been protesting, calling for politicians to take action on combating climate change. Environmental education must be given top priority as we all have become witnesses to the destruction of our ecosystem and a dramatic decrease in biodiversity. Creating awareness is not sufficient; we must find a way to implement change with conviction.

We must think globally, but act locally. This should even apply when electing political leaders. Our voting patterns are also an ethical issue. Today we are witnessing a strong connection between environmental politics and elections.

People have elected greater numbers of Green parliamentarians in Germany, Switzerland, Finland, Belgium, the Netherlands and the European Parliament. This is a strong indication that public opinion and actions can change politicians' minds.

Fortunately, especially young people today understand the connection between environmental politics and elections.

Franz Alt: *What's your response to what Greta Thunberg and young people like her are doing? Do you support* *young students taking time off from school and standing up as activists to demand radical change?*

Dalai Lama: I wrote a letter to her. I really admire what she's doing. We older people will probably be able to manage over the next one or two decades. But the lives of young people like her may extend until the end of this century, and they will have to face whatever changes come about. Therefore, it is quite right that students and today's younger generation should have serious concerns about the climate crisis and its effect on the environment. They are being very realistic. We should encourage them. Sometimes it seems that older people lead a more materialistic way of life; they belong to a more materialistic culture. Younger people are beginning to feel there is something lacking in such a way of life. We should encourage them.

Franz Alt: *Greta Thunberg is quite realistic about politics. She said to members of the US Congress: "Don't invite us here to just tell us how inspiring we are without actually doing anything about it." What can we do about climate change now?*

Dalai Lama: Well, we could do a lot. You come

from Germany. After 1945, European history has shown that peace is possible, even though in Europe in the last century everyone had been at war against one another. I have great admiration for the spirit of the European Union that has preserved peace among its members. No country within the European Union has waged war against another one. Seventy years of peace! The European Union was rightly awarded the Nobel Peace Prize in 2012. Politics can change just as people can. The European Union is a wonderful peace project that gives me great encouragement.

Every crisis creates opportunities. Many people experience this in their private lives. But also crises in politics and the economy always create opportunities. We are always the same people—at all levels.

VI.

The Mountains Here Are as Bald as a Monk's Head

1. Plant trees

Franz Alt: *As a Buddhist monk, you also rely on the power of thought—as already mentioned. Thoughts and their energies are transported further on a spiritual level. Thoughts are energies that form in our minds. Our positive energies can have a positive impact, negative thoughts, of course, a negative effect. What could this Buddhist thinking do for more and better climate protection?*

Looking at the Himalayas from your exile, I recall one of your quotations: "The Himalayan mountains have become as bald as a monk's head." [The Dalai Lama laughs and scratches his bald head.] *Thirty years ago on German television I already showed the brutal deforestation of the*

Tibetan forests by the Chinese. How can we stop the environmental destruction and how can the climate still be saved?

Dalai Lama: Only when we understand that our earth is like a mother, Mother Earth, will we really take care of her. We Tibetans, like the ancient Indian peoples, understand this interdependence: healthy earth, healthy animals, healthy plants, healthy forests, healthy water, healthy people. Mother Earth warns us today: "My children are behaving badly"; she is warning us that there are limits to our actions.

Today we are consuming as much coal, gas, oil and petrol a day that took nature a million years to form. That is the cause of global warming. As a Tibetan Buddhist monk, I am committed to a moderation in our consumption patterns. A responsible life is a simple and contented life. We must learn to cooperate, work and live with nature and not against nature.

Franz Alt: *In Tibet, China has cut down 85 percent of all trees, thus depriving your country of its life force. Why did the Chinese cut down Tibet's forests and what are the consequences for your native county?*

Dalai Lama: When the forests in Tibet die, a whole nation suffers. And when a people suffers, the whole world suffers. We need forests for our health as well. When we go for a walk in a forest, fresh air is healing. We need green forests. They are nature's great gift. Forests are good for our soul. In the forest we find the calm that our brain needs for regeneration. Forests are water reservoirs, home to many animal and plant species, and are important as an air-conditioning machine. They are a mirror of the diversity of life.

The large-scale deforestation in Tibet is a matter of great sadness. It is not only sad for the local area, which has lost its beauty, but for the local people. The deforestation of the Tibetan Plateau, according to experts, will change the amount of reflection from snow into space [forested areas absorb more solar radiation], and this affects the monsoons not only in Tibet, but in all surrounding areas. Therefore, it becomes even more important to conserve Tibet's environment.

The environmental destruction in Tibet clearly shows that Chinese Communist ideology lacks what, in our Tibetan culture, is meant by interdependence or universal responsibility. This also surprises me because communists like to sing "The International" *[again laughing].* Today no nation can solve its problems on its own.

Franz Alt: *Interjection, dear friend! In the United States President Trump rules according to the motto "America first" and "Make America great again." Is this motto still up to date in times of globalization?*

Dalai Lama: When the president says, "America first," he makes his voters happy. I can understand that. But from a global point of view this statement is not relevant. In today's global world, everything is interconnected. America's future also depends on Europe and Europe's future also on Asian countries. The new reality means that everything is related to everything. The USA is the leading nation of the free world. That's why the US president should think more about global-level issues.

Franz Alt: *Should a contemporary topic not be, "Make the planet great again"?*

Dalai Lama: Certainly! The US is still very powerful. The motto of America's ancestors was peace, liberty and democracy. The totalitarian systems have no future. As a leading power the USA should ally itself closely with Europe. I am an admirer of the European Union. The EU is a big and exemplary peace project. Unfortunately, President Trump had announced the US withdrawal from the Paris Agreement. He may have his reasons. But I do not support them. The EU should also become a role model in terms of climate protection. Each and every one should become a climate protector. However, we will not reach this goal through egoism and nationalism, but through the promotion of a sense of oneness of humanity.

Franz Alt: *You suggest planting trees for the future and for peace. Why is that so important?*

Dalai Lama: Trees have been our companions through history and they remain important today. They purify the air for living beings to breathe. Their shade provides a refreshing place to rest and serves as a place for insects and birds to live. They contribute to timely rainfall, which nourishes crops and livestock and balances the climate. They create an attractive landscape, pleasing to the eye and calming for the mind, and continually replenish their surroundings. Properly managed, they are also a source of economic prosperity.

When the environment becomes damaged and polluted, oceans and lakes lose their cool and soothing qualities, so the creatures depending on them are disturbed. The decline of vegetation and forest cover causes the earth's bounty to decline. Rain no longer falls when required, the soil dries and erodes, and forest fires rage. We all suffer the consequences, whether we are ants in the jungle or human beings in cities.

In the context of Buddhism, trees are often mentioned in accounts of the principal events of Buddha's life. He was born as his mother leaned against a tree for support. He attained enlightenment seated beneath a tree, and finally passed away as trees stood witness overhead. According to the monastic code of discipline, fully ordained monks are enjoined not only to avoid cutting trees or grass, but also to plant and nurture them.

Therefore, it is in our own interest to plant trees and flowers around the places where we live, work and study, as well as around hospitals and alongside paths and roads.

In the Tibetan monasteries in Tibet and India, we

have been cultivating tree plantations over the past few decades. This brings in the action of serving others and of creating a better environment and a happier place. [Please also read the Dalai Lama's tree poem in Part VII of this book.] And to truly acquire a sense of responsibility for community, one has to first feel responsible for one's own place or home.

The longing for nature and green is ingrained in us. Human beings love green so much that they plant more and more trees in our cities and towns, and even trees on the roofs. When you spend time in the forest and hear birds singing, you feel good inside. The healing power of forests is becoming increasingly important. When we are surrounded by artificial things, it's harder to be peaceful. It's as if we begin to be artificial; we develop hypocrisy, suspicion and distrust. In that state it's hard to develop genuine, warmhearted friendship. We all feel the need to be surrounded by life. We need life around us that grows, flourishes and thrives. Because, as social animals, we also want to grow, flourish and thrive. We all love our technology. But our relationship with plants and nature is inextricably very old and very deep. Buddhist ethics embrace all life, not solely human but also animal and plant life.

Franz Alt: *The destruction of the environment has now also reached the Tibetan highlands. In the West, many people are still dreaming of Tibet as a paradise, a Shangri-la. Is Tibet still a paradise?*

Dalai Lama: What the Chinese did on the roof of the world after 1959—especially during the Cultural Revolution—was a cultural genocide. What I learn today from Tibetan refugees who I meet in my exile in Dharamsala makes me fear that my home country is rather the opposite of a paradise. But I find admirable how the vast majority of Tibetans, seventy years after the occupation, still cling to their religion, language and culture, and respect for environment even though there are more Chinese than Tibetans today living in Lhasa, Tibet's capital. The Chinese have reduced us to a minority in our own country.

Franz Alt: *Do you think that you will reach this goal and return to Tibet someday?*

Dalai Lama: China is a great nation, an ancient nation—but its political system is a totalitarian system, no freedom. I am happy to live in India for the rest of my life. I can live in this country and utilize the Indian freedom to fulfill my commitments toward promotion of human values, religious harmony, protection of Tibetan culture and environment, and revival of ancient Indian knowledge.

Franz Alt: *Should the Buddha return to our world and join a political party, he would certainly be Green, you say. What makes you so sure?*

Dalai Lama: Buddha and we Buddhists have deep respect for nature and evolution. We know that nature does not need us humans, but we do need nature. Looking at today's global exploitation of nature, I think: without humans the earth would do better *[again laughing]*.

Franz Alt: *Which political party would you support?*

Dalai Lama: I have no hesitation in supporting initiatives that are related to protecting the environment. In Europe I would vote for the Green Party, because threats to our environment are a question of our survival. This beautiful blue planet is our only home. It provides a habitat for unique and diverse communities. Taking care of our planet is to look after our own home.

2. Ethics is more important than religion

Franz Alt: *And you would vote Green if you lived in a Western democracy. Why?*

Dalai Lama: Because they represent a similar nature-friendly philosophy as we Buddhists do. For over a thousand years, nature has been sacred to us Tibetans. On the high Himalayan plateau where we live, we try, in the spirit of Buddhism, to live in peace with nature, protected by our mountains, without violence and in compassion with all living beings. Nature is sacred to us. Nature is our true home. We humans come from nature. We can live without religion, but not without nature. Therefore, I say that environmental ethics are more important than religion. If we keep destroying nature as we are doing today, we will not survive.

This is a law of nature that we have to accept. Mankind will suffer terribly if we do not learn that: a clean environment is a human right like any other human rights. It is our responsibility toward all sentient beings to ensure that we leave to our children and grandchildren a world at least as intact as we found it when we were born. There are limits to what we are allowed to do, but no limits to our universal responsibility.

Franz Alt: *What do you do yourself for the environment and for the climate?*

Dalai Lama: On a personal and family level, too, we need to develop a much clearer awareness of our actions and their consequences, such as how we use water or dispose of our garbage, so that taking care of and limiting damage to the natural environment becomes an ordinary part of our daily lifestyle. That is the proper way, and it can only be achieved through education.

I switch off the light when I leave my room. I take a shower instead of a bath. I eat little meat. I encourage other people to do the same. We must think globally, but act locally. This should even apply when electing political leaders. Our voting patterns are also an ethical issue. We should all vote for the real environmentalists. Fortunately, especially young people today understand the connection between environmental politics and elections.

As someone born in Tibet, the rooftop of the world, where Asia's great rivers start and the world's highest peaks are to be found, I have loved nature since my childhood. I have made environmental conservation one of my life's commitments and advocate protection of the environment wherever I go.

I honor my promise to Mr. Sunderlal Bahuguna, an Indian environmentalist, to speak about environment preservation. When I travel to the trans-Himalayan region from Ladakh to Arunachal Pradesh in India, I urge people there to plant trees to save their land from becoming barren in future. Trees make the landscape's greenery and bring peace and happiness of mind in our day-to-day life.

3. Vegetarianism helps the climate

Franz Alt: *In 1965 you became a vegetarian. Have you been vegetarian since then and why?*

Dalai Lama: In 1965, I became completely vegetarian—no eggs, nothing. But instead, I was eating lots of cream and nuts, and after twenty months, I had trouble with my gallbladder and got jaundice. My skin, eyes, nails—everything—turned yellow. My doctors advised me that I should go back to my original diet. I should again eat some meat, which I am doing about once or twice a week. So, I am a little bit of a contradiction, telling people to be vegetarian as a non-vegetarian myself.

Nevertheless, right from the beginning, from the time when I was in Tibet, I worked very hard to promote vegetarianism in Tibetan society. In the late 1940s, all the food served during Tibet's official festivals used to be vegetarian. Even campaigns promoting vegetarianism have been launched in the communities. In India most of the Tibetan monastic institution's kitchens have now started serving only vegetarian food to their monks and nuns.

Franz Alt: *Worldwide meat consumption is rising. So more and more billions of animals have to be killed. Do you think that worldwide meat consumption could drop again and how is that to be achieved?*

Dalai Lama: Buddhism does not forbid eating meat. But it is a question of how and how much. Buddhism says no animals should be killed for eating. But our attitude toward meat is rather curious. Tibetan Buddhists can buy meat but should not kill animals.

What I find particularly worrying is intensive animal husbandry. We humans can live largely without or with little meat. And above all without animal suffering—in particular in our modern world, where we have many alternatives, especially fruits and vegetables. In the meantime, there is even meat made from vegetables, for instance from peas and beetroot, from potatoes and coconuts.

Intensive animal husbandry has serious consequences not only for animals, but also for man's health, the soil, insects and the air.

4. Buddhists disapprove of killing as a sport

Franz Alt: *What you call "curious" is also the attitude of most Western Christians toward meat. Me too, I am only an 85 percent vegetarian. My doctor also recommends eating meat or fish about once a week for health reasons. If we had to slaughter ourselves the animals we eat, most of us would probably be strict vegetarians. That is also rather curious.*

Meat consumption and animal husbandry cause about as many greenhouse gases globally as all cars, planes, trains and ships put together. Besides, environmental physician Professor Hans-Peter Hüttner of the Medical University of

Vienna says: "Meat plays a crucial role in the development of cancer of the intestine or circulatory diseases. Moderate meat consumption significantly reduces your disease risk, is beneficial to environment and climate and doesn't really hurt. A win-win situation." What I eat affects all of us. What do you think about hunting and fishing as a sport?

Dalai Lama: We Buddhists disapprove of killing as a sport. I support those groups and people who work for animal rights and animal welfare around the world. It is sad that millions and billions of animals are killed for human consumption.

I once visited a poultry farm in Japan with 200,000 hens. They were kept in small cages just to produce eggs for two years. After that they were sold for slaughter. That was shocking. We should support people who fight against such unworthy business and such animal misery. It is also very dangerous and shortsighted for us to simply suppress and forget about animal suffering. What we are doing to animals today, can also happen to us. Perhaps one day we will kneel down and ask the animals for forgiveness. I also disapprove of the way in which we have mechanized farming these days.

We must never forget the suffering we inflict on

other sentient beings. I am thinking of some Tibetan butchers or Japanese fishermen who ask the animals they kill for forgiveness and pray for them.

Franz Alt: *Do you think that worldwide meat consumption could drop again, and how is that to be achieved?*

Dalai Lama: In some countries this is already the case. I meet young people everywhere who are looking for alternatives to brutal meat consumption. The new meatless brand in the USA is called "Beyond Meat." Many consumers want to reduce meat consumption in order to protect the climate, but also to alleviate animal suffering caused by factory farming. Now there are vegetarian "hamburgers" *[laughing]*.

Franz Alt: *But that means that we Western people would have to learn to let go at least a little bit? Is letting go the heart of greater environmental justice?*

Dalai Lama: Yes, you can say so. Letting go of surplus is the heart of spiritual growth. Just imagine what we could achieve if the USA only halved its military budget. That would be over $300 billion every year for environmental projects such as the solar energy transition or overcoming hunger in poor countries. Defending the future instead of dangerous military upgrading. That could indeed be the beginning and the energy for an ecological age. "Letting go" would mean liberation.

Franz Alt: *Leo Tolstoy said: "As long as there are slaughterhouses, there will be battlefields." Can you agree with this? Do you also see this connection between intensive livestock farming and violence among people?*

Dalai Lama: Such connection exists. In all religions we know this spiritual law: "As you sow, so shall you reap." By nature we have inner inhibitions about killing. Above all, we feel that it is not right to inflict pain on other sentient beings. If we let our conscience be brutalized while killing animals, it will also be brutalized while killing people.

Franz Alt: *"Whoever changes, changes the world," you say. Western societies are mainly concerned with today's generations. Former generations do not exist anymore. And the future ones do not exist yet. And in the West only a minority believe in rebirth. So what chance do environmental and climate protection have in this situation?*

Dalai Lama: As I already said: the belief in reincarnation can help to protect the environment. We must not leave the young people alone in their fight for a good environment and climate. It is important that we now organize and publicize and have global solidarity with the young people. Everybody and everything must change if we want to live in a climate that is compatible with life. Our generation has damaged the climate, so we also should help to save it.

Franz Alt: *For many months now, hundreds of thousands of young people have been demonstrating for more and better climate protection in over 100 countries. Do these young demonstrators give you hope?*

Dalai Lama: Members of the younger generation, to whom the twenty-first century belongs, have important responsibilities: they must learn from and rectify the mistakes of the past, and ensure that such mistakes are not repeated. The younger generation, who will inherit this earth, has the ability and the opportunity to act and create a more compassionate world. The twentieth century experienced immense destruction, human suffering and unprecedented environmental damage. I urge them to make the twenty-first century a century of dialogue and a century of compassion, including on the issue of environment. We need a revolution of compassion based on warmheartedness, a sense of concern for others' well-being, and respect for the rights of others.

Franz Alt: *Your Holiness, dear friend. I cordially thank you for these reflections, which we have been exchanging for decades. They will help many people to understand that our twenty-first century must become the one in which globalized mankind finds ways to universal responsibility. I have learned from you that every single one of us can and must take on their own piece of universal responsibility if we want a better world. Inner peace, love and compassion are the most important energies that will lead to external peace as well. And to peace with nature.*

Whenever the Dalai Lama and I say goodbye, he places a *kata*, a greeting scarf made of white silk with the traditional sign of good luck, around my neck. He takes my head between his hands, our foreheads and noses touch one another in friendship, and we embrace for a long time and sense that love and peace between us humans is possible. It is the spirit that gives strength to peace, justice and friendship.

VII.

The Sheltering Tree of Interdependence—A Buddhist Monk's Reflections on Ecological Responsibility

This poem was released on the occasion of the presentation by His Holiness the Dalai Lama of a statue of the Buddha to the people of India and to mark the opening of the International Conference on Ecological Responsibility: A Dialogue with Buddhism on October 2, 1993, in New Delhi. (A booklet of the poem, in Tibetan and English, is distributed by Tibet House, New Delhi.)

During the course of my extensive travels to countries across the world, rich and poor, East and West, I have seen people reveling in pleasure, and people suffering. The advancement of science and technology seems to have achieved little more than linear, numerical improvement: development often means little more than more mansions in more cities. As a result, the ecological balance—the very basis of our life on earth—has been greatly affected.

On the other hand, in days gone by, the people of Tibet lived a happy life, untroubled by pollution, in natural conditions. Today, all over the world, including Tibet, ecological degradation is fast overtaking us. I am wholly convinced that, if all of us do not make a concerted effort, with a sense of universal responsibility, we will see the gradual breakdown of the fragile ecosystems that support us, resulting in an irreversible and irrevocable degradation of our planet, earth.

These stanzas have been composed to underline my deep concern, and to call upon all concerned people to make continuous efforts to reverse and remedy the degradation of our environment.

1. O Lord Tathagata
   Born of the lksvakus tree
   Peerless One
   Who, seeing the all-pervasive nature
   Of interdependence
   Between the Environment and sentient beings

Samsara and Nirvana
Moving and unmoving
Teaches the world out of compassion
Bestow thy benevolence on us

2. O the Savior
The one called Avalokitesvara
Personifying the body of compassion
Of all Buddhas
We beseech thee to make our spirits ripen
And fructify to observe reality
Bereft of illusion

3. Our obdurate egocentricity
Ingrained in our minds
Since beginningless time
Contaminates, defiles and pollutes
The environment
Created by the common karma
Of all sentient beings

4. Lakes and ponds have lost
Their clarity, their coolness
The atmosphere is poisoned
Nature's celestial canopy in the fiery firmament

Has burst asunder
And sentient beings suffer diseases
Unknown before

5. Perennial Snow Mountains resplendent in their glory
Bow down and melt into water
The majestic oceans lose their ageless equilibrium
And inundate islands

6. The dangers of fire, water and wind are limitless
Sweltering heat dries up our lush forests
Lashing our world with unprecedented storms
And the oceans surrender their salt to the elements

7. Though people lack not wealth
They cannot afford to breathe clean air
Rains and streams cleanse not
But remain inert and powerless liquids

8. Human beings
And countless beings

That inhabit water and land
Reel under the yoke of physical pain
Caused by malevolent diseases
Their minds are dulled
With sloth, stupor and ignorance
The joys of the body and spirit
Are far, far away

9. We needlessly pollute
The fair bosom of our mother earth
Rip out her trees to feed our shortsighted greed
Turning our fertile earth into a sterile desert

10. The interdependent nature
Of the external environment
And people's inward nature
Described in tantras
Works on Medicine, and astronomy
Has verily been vindicated
By our present experience

11. The earth is home to living beings;
Equal and impartial to the moving and unmoving

Thus spoke the Buddha in truthful voice
With the great earth for witness

12. As a noble being recognizes the kindness
Of a sentient mother
And makes recompense for it
So the earth, the universal mother
Which nurtures equally
Should be regarded with affection and care

13. Forsake wastage
Pollution not the clean, clear nature
Of the four elements
And destroy the well-being of people
But absorb yourself in actions
That are beneficial to all

14. Under a tree was the great Saga Buddha born
Under a tree, he overcame passion
And attained enlightenment
Under two trees did he pass in Nirvana
Verily, the Buddha held the tree in great esteem

15. Here, where Manjushri's emanation
Lama Tsongkhapa's body bloomed forth

Is marked by a sandal tree
Bearing a hundred thousand images of the Buddha

16. Is it not well known
That some transcendental deities
Eminent local deities and spirits
Make their abode in trees?

17. Flourishing trees clean the wind
Help us breathe the sustaining air of life
They please the eye and sooth the mind
Their shade makes a welcome resting place

18. In Vinaya, the Buddha taught monks
To care for tender trees
From this, we learn the virtue
Of planting, of nurturing trees

19. The Buddha forbade monks to cut
Cause others to cut living plants
Destroy seeds or defile the fresh green grass
Should this not inspire us
To love and protect our environment?

20. They say, in the celestial realms
The trees emanate
The Buddha's blessings
And echo the sound
Of basic Buddhist doctrines
Like impermanence

21. It is trees that bring rain
Trees that hold the essence of the soil
Kalpa-Taru, the tree of wish fulfillment
Virtually resides on earth
To serve all purposes

22. In times of yore
Our forbears ate the fruits of trees
Wore their leaves
Discovered fire by the attrition of wood
Took refuge amidst the foliage of trees
When they encountered danger

23. Even in this age of science
Of technology
Trees provide us shelter
The chairs we sit in

The beds we lie on
When the heart is ablaze
With the fire of anger
Fueled by wrangling
Trees bring refreshing, welcome coolness

24. In the trees lie the roars
Of all life on earth
When it vanishes
The land exemplified by the name
Of the Jambu tree
Will remain no more than a dreary, desolate desert

25. Nothing is dearer to the living than life
Recognizing this, in the Vinaya rules
The Buddha lays down prohibitions
Like the use of water with living creatures

26. In the remoteness of the Himalayas
In the days of yore, the land of Tibet
Observed a ban on hunting, on fishing
And, during designated periods, even construction

These traditions are noble
For they preserve and cherish
The lives of humble; helpless, defenseless creatures

27. Playing with the lives of other beings
Without sensitivity or hesitation
As in the act of hunting or fishing for sport
Is an act of heedless, needless violence
A violation of the solemn rights
Of all living beings

28. Being attentive to the nature
Of interdependence of all creatures
Both animate and inanimate
One should never slacken in one's efforts
To preserve and conserve nature's energy

29. On a certain day, month and year
One should observe the ceremony of tree planting
Thus, one fulfills one's responsibilities
Serves one's fellow beings
Which not only brings one happiness
But benefits all

30. May the force of observing that which is right
    And abstinence from wrong practices and evil deeds
    Nourish and augment the prosperity of the world,
    May it invigorate living beings and help them blossom
    May sylvan joy and pristine happiness
    Ever increase, ever spread and encompass all that is

VIII.

For a Solar Age—
Epilogue by Franz Alt

1. Reconciling economy and ecology

Climate change is not distant—it's already here. We have to face reality, which may be complicated, but not hopeless. Since the Enlightenment about 300 years ago, at the latest, and the "de-idealization" of the world, philosophy has no longer been "Theology's handmaid" *("ancilla theologiae")* and our Western cultural model is based on scientific knowledge—at least we think so.

But why have scientists worldwide been warning for decades against climate change without actually being heard in politics and society, let alone bringing about appropriate actions? Why is the Enlightenment of the past not sufficient to secure our salvation? In order to prevent the worst, we need a new Enlightenment, a second one, more profound, an "enlightening" of the Enlightenment.

In this book, the Dalai Lama demonstrates very clearly that today's environmental crisis is the crisis of our inner world.

We think that we know what we are doing. But actually, we are not doing what we know. The idea that rationalism alone will save us is quite irrational. Reason alone will not bring man to his senses. We love to suppress such insight. There are those, however, that are fully aware of what they are doing.

And many are afraid of the necessary changes. It is true that sometimes politicians opt for necessary changes out of fear. So the CFC ban resulted from the fear of skin cancer. The introduction of the three-way catalytic converter in German cars resulted from the fear of dying forests.

The lobbyists of the traditional energy business are simply afraid of losing their benefits. Politicians also know about their reliance on the greed of big companies, and many citizens travel by plane or by car and eat themselves sick with a lot of meat, although they know what they are doing to themselves, the environment and their children's future. Can humans really confront this reality? Is the classical, purely rationalistic Enlightenment sufficient for us?

Conservatives and religious people have trusted the wisdom of nature for millennia. But this confidence is being shaken in times of global warming and the extinction of species. Mother Nature does not cooperate anymore but is running riot; she has got a fever and is on strike. We are about to lose our partner and confidante—nature, the source of our wealth and happiness. But first we should at least understand that we do not want to save the climate as such, but—quite egotistically—ourselves.

And now, by means of a second Enlightenment, we will have to achieve success in thinking that combines religion and philosophy, nature and reason, freedom and responsibility. This is what we call eco-spirituality in this book. Here the Dalai Lama also speaks of education of the heart. If we do not understand this interdependence, our liberty will soon end in enslavement. Global warming is already creating a lack of freedom, for instance for refugees or farmers who are losing their land to the desert or for elderly people who died in Europe during the hot summers of 2003 and 2018. In 2003 alone—according to EU statistics—about 60,000 people died of the heat within the EU. In India, where temperatures rose to unbearable 122 degrees Fahrenheit, hundreds of thousands of elderly people died from overheating. It is absolutely surprising to see how difficult it is, especially for conservatives, to understand these correlations, whereas helping to preserve the creation should be the conservatives' central task. Ah, I wish the conservatives were really conservative!

After the Enlightenment, many intellectuals believed that we could become emancipated not only from our self-inflicted immaturity but also from nature. Within only 300 years we robbed nature of her resources that she had gathered in 300 million years, blew the waste into the air and filled the gigantic holes that were made with gigantic waste. Today's economists call it progress. What now?

2. There is no matter

For 300 years economists have believed that money was the basis of all economizing. Nature, however, is the basis of an economy that deserves this name. Economizing on dead soil makes little sense and does not need any jobs.

Both economy and ecology derive from the Greek *oikos*, i.e., economizing. The power of money is the principal disease of our time. "The attempt to separate numbers from values results in total dominance of the numbers," writes Christian Felber, Austrian economist, author and founder of the "welfare economy." And: "Separating economy from ecology is one of the greatest sins committed by economists."

The increase of money does not make the world richer. In the past fifty years we have brought about the most brutal kind of poverty by killing off about half the species of animal and plant life. More money, but less richness of life: unlike this crazy economy, theology is almost an exact science. Today's economists should learn to place their science in a broader holistic context. Pope Francis says: "Everything is connected with everything: the poor and nature are crying for help."

Joachim Schellnhuber, climatologist and physicist, has reached a similar conclusion and summarizes it as follows: "Speaking to an economist is the maximum penalty for a physicist."

There will be no future unless we learn presently to avoid the biggest mistakes of the past. Only then will we arrive in the real world. Or in the divine order.

In religious language, God is identical with spirit, according to the Gospel of John. With this in mind, Nobel laureate Max Planck says: "As a physicist, i.e. as somebody who has devoted his whole life to the most clear-headed science, to the study of matter, I will certainly not be held to be a dreamer. I can tell you as a result of my research about atoms this much: there is no matter as such. All matter originates and exists only by virtue of a force which brings the particle of an atom to vibration and holds this most minute solar system of the atom together. We must assume behind this force the existence of a conscious and intelligent Mind. Not the visible yet transient matter is real and true, but rather the invisible and immortal spirit. As there cannot be a mind as such and as every mind belongs to a being, we are forced to assume the existence of spiritual beings. As these beings cannot exist by themselves but must be created, I have no hesitation in calling this mysterious Creator, like all ancient civilizations—God!"

Professor Hans-Peter Dürr, physicist and former Director at the Max Planck Institute in Munich, argues similarly in his popular book *There Is No Matter.* Shortly before he died, we walked up to the Acropolis in Athens. He already had great difficulty breathing. Surrounded by so much matter of stones and rocks, I said to the world-renowned physicist: "Everything here is actually matter." "Oh no," he replied—to the dismay of many a Marxist and materialist: "All this is materialized spirit. The spirit has been, and will always be, primary." A surprising insight for a physicist.

3. In depth all life is one

Hans-Peter Dürr's research shows amazing parallels between Judaic-Christian thinking, Hindu-Buddhist insights and the latest findings of modern quantum physics. We must finally learn to cross borders in order to overcome what seems irreconcilable. The borders of pure rationalism and the Enlightenment run at the surface—IN DEPTH ALL LIFE IS ONE.

This is the conclusion of both today's quantum physicists and the Dalai Lama, according to the German physicist Carl Friedrich von Weizsäcker.

If we understand, or imagine, God as the sun behind the sun, solar thinking and acting will be more than the purely technical transition from the fossil age to a solar age; the transition is a sign of a new, profound, holistic and mature attitude toward life of divine substance. Tens of thousands of people worldwide inspired with this attitude and change of mind tell me: "We have now taken a new attitude toward nature and the sun—we're looking upward more often and also understand what Jesus meant in the Sermon on the Mount, saying: 'He causes his sun to rise on the evil and the good.'" There is simply no sun only for energy companies. The sun, a divine symbol, shines on all of us, a master craftsman creating inexhaustible masterpieces and inspiring primarily the wonders of the mind.

4. No child should starve to death

To many "solar humans," looking upward also means looking inside. They learn to gain access to their souls through their dreams. Due to their external energy autocracy, they become aware of their inner energy sovereignty. They recognize that the outer energy crisis corresponds to a much deeper inner energy crisis, an energy crisis of their soul. And the human soul, according to the renowned Swiss depth psychologist Carl Gustav Jung, "is the only superpower of this world that I recognize."

If we acted in a way that is truly holistic and based on knowledge, with an awareness of our soul, we would have reduced the greenhouse gases a long time ago and not only articulated but also realized transition in the field of energy and transport, water and agriculture. But we are living in a period that, to a great extent, is unaware of our soul, which makes us resistant to, and incapable of, learning. It seems that only inner healing can bring about outer healing. As the mystics of the Middle Ages already knew—outer like inner, inner like outer.

Our "inner light" (as described by Jesus, in the Sermon on the Mount), our soul can help us to become a sunny being to the outside. And we may suddenly sense that the most important things in life are free: joy and gratitude in ourselves, as well as the sun above us. Yet joy, love and gratitude are as little for sale as the sun. They are God's gift. Let us finally make intelligent use of them. For this purpose we should open our hearts in the same way as we can place the roofs and walls of our houses at disposal for solar installations, as landing strips for the spirit from above. So we can—perhaps for the first time in mankind's history—work together so that one day children will no longer have to starve and people will not be forced to flee their country. And we will learn that nobody leaves his home voluntarily. Energy crisis, refugee crisis and climate crisis are closely linked. If we see this correlation, we will find solutions; crises give rise to new opportunities. The key to the solution of all these crises is the energy crisis.

In October 2018, I was invited to give lectures in Mali/Western Africa. In Bamako, Mali's capital, I was expected to speak at an African solar energy conference. Africa and the sun: what a chance! Upon invitation of the Minister of Energy, we speakers traveled to a village of 20,000 inhabitants, outside the capital. Only three years ago, there had been no electricity there. But meanwhile people now had solar installations and electricity. The village pharmacist told me that by means of solar energy, the health level of the village had risen. He now could—unlike before—cool a lot of medicine. In a school, I got to know children who enthused about the benefits of solar energy, because now they would watch soccer on TV. Mothers report that now they can send their children to school, as they can do their homework in the evening, in the light of solar lamps, which was impossible before. Education changes everything. A dressmaker told me that at last she has an electric sewing machine and did not have to slave away at the treadle any more. The village mayor said to me: "Since we've had electricity in the village, young people don't think of fleeing to Europe any more. Solar energy also creates new jobs."

All problems created by man can also be solved by man.

Shortly afterward I gave a lecture at the Conference of the World Wind Energy Association in Karachi/Pakistan. Here the Chinese company Goldwind, one of the world's greatest producers of wind turbines, has—with the help of German technology—set up a wind park that produces affordable and clean electricity for 1.5 million people in Karachi. My Pakistani friends are full of praise for such progress. They are fully convinced that their country will achieve the complete energy transition by the middle of this century. Karachi and Mali could be everywhere.

Every wind turbine and every solar installation, every hydropower plant and every biogas plant is a sign of peace. Never are wars waged over the sun or the wind.

5. Disarming instead of rearming

Can this trend be financed at all? This is a typical and immediate objection, often raised in Germany. Today we spend about 1.6 trillion dollars worldwide on weapons and the military, in continued accordance with the principle of ancient Rome: "If you want peace, you must prepare for war." The result of this traditional way of thinking: 2,000 years of war, misery, destruction and annihilation. Millions of serious casualties. In the atomic age it is now high time we reversed this motto: "If you want peace, you must prepare for peace." This means disarming instead of rearming. Solar energy transition can be financed worldwide with just a small fraction of the money spent on war preparations. We are still caught in the old war trap.

So what are we waiting for? We humans can learn and change our ways. As early as in 1972, the much-discussed book *The Limits to Growth* of the Club of Rome was published. At that time we produced a special broadcast on this topic in our programme "Report Baden-Baden" and reached an audience of millions. And for quite some time the newspapers covered this crucial topic in detail. The book became a bestseller and was translated into all world languages. We have known for a long time that on this earth all matter is limited. We all grow until we are 160, 170, 180 or 190, or perhaps some more, centimeters tall. But then it's enough. Nobody grows physically forever. After physical growth something else matters: inner maturity, inner growth.

As far as spirit, mind, culture or religion is concerned, we can continue to mature, but can never grow physically forever. Our material resources are limited, unlike the ideas of our mind. Ideas can multiply as if they had sexually reproduced. This is the basis of progress and prosperity. Nevertheless, governments worldwide are propagating economic growth. The only material being that grows without limits is cancer. Our philosophy of eternal growth is propagating a quasi-official cancerous economy, and is doing so worldwide. This could prove fatal.

But how can we, in times of global warming and environmental destruction, grow spiritually so that we will be up to these challenges, before it is too late? How can we mature rather than grow? This may be the question of all current questions.

The greatest lie in politics: "But we are already doing so much." In my TV series *Zeitsprung*, we showed that in 1993, mankind was blowing about twenty-two billion tons of greenhouse gases annually into the atmosphere. And as if this were not enough, it is today almost forty billion tons per year—after more than twenty world climate conferences with tens of thousands of participants, more than twenty-seven years later.

But we are already doing so much, aren't we? Yes we are, but it's all wrong!

Every day we are emitting globally about 150 million tons of CO_2, every day we are killing off about 150 animal and plant species, we are losing daily

50,000 tons of fertile soil and deserts grow by about 80,000 hectares every day. Our greed for meat destroys rain forests. In 2019, hundreds of millions of Africans lived through the most terrible drought within living memory and are fearful of the next famine. A whole subcontinent is crying for water: Angola, Botswana, Congo, Lesotho, Malawi, Mozambique, Namibia, Rwanda, Zambia, Zimbabwe, South Africa. Is there still hope for us?

6. Economizing with nature, not against it

We are experiencing not ecological standstill but a speedy step back. We are racing toward the abyss, with politicians cheering us on: "Speed up! It's the only way to save us. Therefore: growth, growth, growth." This is perverse in the extreme. But how can we turn the corner before everything keels over and we're finally racing into the abyss?

We are destroying our earth because we are insanely concerned with the present and the future. The motto of our Western cultural model continues: greater always means better. The fact that more and more ever-larger SUVs are being bought in Germany is a case in point. In our cities an increasing number of these vehicles weighing 2.5 tons and measuring 5.2 by 2 yards are running against the world climate, although space in the cities is getting tighter and tighter. Owners and drivers, women and men alike, are more concerned with their ego than with mobility.

Most of us couldn't care less about the future. Having just concluded my speech on energy transition, I had this experience: an elderly gentleman came to the book table, saying: "Well, Mister Alt, you may be right about solar and wind energy, but, you know: I'm seventy-five by now, there's enough for me." "Have you got children?" I retorted. He bowed his head, dumbfounded, and went away.

A widespread motto seems to be: "After us, the deluge." We may be the first generation who can no longer say to their children in all conscience: "We love you." Many children would have to answer their parents: "We don't believe you. This is hypocritical. You're only pretending. If you really loved us, you wouldn't burn our future."

Many children and young people begin to see through us and revolt against our pyromania. Immediate solar energy transition has become a matter of survival for mankind. Fortunately, as many as 28,000 climatologists from "Scientists for Future" are supporting the "Fridays for Future" movement, as well as "Parents for Future," even "Grandparents for Future," "Farmers for Future," "Doctors for Future," "Entrepreneurs for Future," not forgetting the first "Journalists for Future" and "Churches for Future," and even "Climbers for Future." This is entirely consistent with the Dalai Lama's position. And absolutely in the spirit of the Pope's Encyclical *"Laudato si."* What the world now needs is the movement "Citizens for Future." The crucial question today is: How can real change be brought about? Not through fear but with positive emotions like compassion and mindfulness.

Without energy, economic development is impossible in countries that are still poor. Without energy, people in emerging countries have no option but to flee to economically rich countries. What would we do if we lived in poorer countries and saw no prospects for our children?

According to a UN prognosis, there will be more than 400 million climate refugees by the end of our century. Of course they are fleeing to countries where they see economic prospects. We now reap what we have sown. We, the industrialized countries, have caused climate change, not the poor countries. The poor are victims of our actions. And that is why they will come here, unless we stop global warming. A Bangladeshi or a sub-Saharan African consumes about twenty times less energy than a German. We are responsible, not the Africans who want to, or have to, come to Europe. Where else should they go?

On average, a US citizen emits eighteen tons of carbon dioxide per year, a German nine tons, a Swede 4.5 and a Bangladeshi or sub-Saharan African 0.5 tons.

So far, mostly war refugees have come over here. In 2015, for instance, from Syria or Afghanistan. Usually they return after the wars to rebuild their home countries, like the refugees from former Yugoslavia, after the wars in the nineties. But where should future climate refugees return to?

Global warming concerns us all. It is a problem that once again we are about to bury. But everything we repress will one day come home to roost. Millions of climate refugees will in the future push their way to Germany, unless we finally put an end to the cause of their flight, global warming. Seriously and immediately.

Climate change is—as already mentioned in the beginning—a world war against nature and concerns all countries. We will have to learn to distinguish between war refugees and those whom climate change forced to flee. A flight of no return.

The history of mankind is a history of refugees. In a way we are all refugees. This began about 200,000 years ago when Homo sapiens left Eastern Africa to conquer the whole wide world.

I am writing these lines on board Germany's ICE, a high-speed train, on the two hundred fiftieth anniversary of the German scholar Alexander von Humboldt, on September 14, 2019. Traveling in Latin America from 1799 to 1804 for the purpose of research and discovery, he gained a myriad of new insights into the laws and richness of nature. Never before had reports on the tropics attracted so much worldwide attention. On the occasion of September 14, many German newspapers call Alexander von Humboldt the "world's first environmentalist." He actually enthused about the "wonders of the lush primeval forests" and their biodiversity. He was "out of his senses," he wrote excitedly. This brilliant and world-famous scientist, always eager to learn, would today also be "out of his senses," if he had to witness the unbelievable and disgusting brutality that is destroying these "wonders of nature"—out of greed and ignorance.

As a universalist and naturalist, he saw the world as a whole. Nowadays, when the findings of scientists who think and do research on a holistic basis are brazenly denied, Humboldt could serve as a model. What he considered to be holistic is called interdependence by the Dalai Lama in this book.

Humboldt's enthusiasm for flowers and leaves, for the rivers and flies of the tropical rain forests, together with scientific meticulousness is contrary to where the spirit of maximal exploitation and brutal greed has led us. This evil zeitgeist has set Amazonia ablaze. In autumn 2019, so much forest in Sumatra was burning that entire cities almost suffocated, thousands of schools had to close, and the blaze released 360 million tons of carbon dioxide in five weeks. The glacial ice is melting in Greenland as well as in Alaska, in the Arctic and Antarctic, in the Alps and in the Himalayas. Permafrost in Siberia is thawing. In the past few decades we have destroyed half of the rain forests, the lungs of our planet. Only one of its lobes has survived.

Along with dying primeval forests, the idea of animated nature, where everything is interrelated, also dies. Thus we are destroying the basis of our own life. Today's US president and the German political party AfD as a whole deny global warming, contrary to all experts. Alexander von Humboldt never distinguished between knowing and sensing, between feeling and reason, between man and nature, between economy and ecology. He knew that man is dependent on the climate. Humboldt's cross-linked knowledge was always coupled with a sense of responsibility for the world as a whole. Alexander von Humboldt, as historian Kia Vahland wrote in the *Süddeutsche Zeitung*, embodies an "energetic universalism, abounding in knowledge," an exemplary model in our time.

Global warming and the threatening worldwide climate catastrophe are the greatest and most important construction sites of the entire twenty-first century and probably far beyond. Global warming affects everybody in every country. It seems that for the first time in the history of mankind, we all have a common enemy. This challenge could, and should, unite us in the fight against the greatest of our enemies, global warming.

It is not overstated, but simply true: the end of our civilization has become possible. And therefore, on the occasion of the world strike day on September 20, two hundred international media organizations united in autumn 2019 for the first time to inform people about this global threat. So nobody can pretend any longer he didn't know.

On September 20, 2019, the Dalai Lama wrote: "Young people worldwide are demonstrating for a good climate, which is wonderful. Thereby they demonstrate a realistic view of their own future. We adults should support these young people." Greta's opponents accuse her of sentimentality and irrationality, although science and reason are on her side, which does not go for her detractors. Naturally, global warming is a highly emotional issue. In her speech at the UN, Thunberg also displayed tears, anger and despair. Such means will of course wear out eventually. But Greta's opponents argue far less rationally.

7. Deeds are evidence of the truth

Transformation is possible—there are always alternatives, as already mentioned in this book. We can do something or we can leave it.

"There's nothing I can do about it." This is the most fateful and fatalistic excuse when people can't think of anything better. However, it is the most widely used excuse.

Everybody is by nature capable of transformation. This is the reason why we are here. All problems caused by man can also be solved by man.

- peace is possible,
- love is possible,
- justice is possible,
- compassion is possible,
- climate protection is possible,
- sustainable economy is possible,
- a better world is possible

Fine words must of course be followed by appropriate deeds. Deeds are the evidence of our truthfulness. Only then will utopias become concrete and realizable visions.

Greed for money is born in ignorance, according to Buddha and Jesus. Greed, it follows, is irrational. No amount of money, no share price, no gross national product, no property can ever be enough to satisfy our greed and end this irrational pursuit. People living in the rich industrialized countries, whose income has more or less doubled every twenty years since 1945, are not happier than before 1945.

The only antidote to money and greed is compassion, as it is taught by Buddhism and early Christianity, as long as it followed Jesus. At their origin, both religions ignore dogmatism; they are committed to pragmatism and science. If science invalidates the scriptures, they have to be rewritten, even the so-called Holy Scriptures.

Masterpieces of literature and, for that matter, fairy tales are imbued with the idea of man's ability to change: *The Divine Comedy* by Dante, Goethe's *Faust*, Homer's *Odyssey*, the Gilgamesh epic *Parzival*, Jesus's "Sermon on the Mount," Plato's "Cave Parable" or Mozart's *The Magic Flute*.

Thanks to modern, twentieth-century psychology, as represented by Sigmund Freud and Carl Gustav Jung, as well as early neurosciences of the twenty-first century—neuropsychology, neurophilosophy and neurobiology—we have recognized that man in principle is able to change and to alter. Such processes of alteration are called "individuation" or "self-actualization" by the Swiss depth psychologist Jung. According to him, individuation means "anima-integration" to the man who tries to integrate the female parts of his soul, and "animus-integration" to the woman integrating the male parts of her soul. For Jung, the "self" or "individuation" stands for "unity and wholeness of the complete personality." Alteration—or, in religious terms, "conversion"—is, in principle, always possible. People can learn if only they want to. Our will may often be blind, but not stupid. We can train it like a muscle. This is why, in the course of history, what had seemed impossible has often become possible: abolition of slavery and child labor, women's emancipation and the separation of state and church, human rights and democracy and in 1989, the German reunification.

IX.

Ten Commandments for the Climate

- By 2035 at the latest, greenhouse gases must be reduced to zero. The most effective way to protect the climate is immediate abandonment of coal-fired power. Slovakia plans to phase out coal by 2030, Greece by 2028 and England, a traditional "coal country," as early as 2025. Why Germany—or the US and UK—not before 2038?

- All new buildings must be emission-free, for instance, by means of more timber constructions. Aluminium as a building material damages the climate 128 times more than wood. More and more Europeans are using wood for their houses.

- From now on, the construction of power plants should only be authorized if renewable energies are used. Cut today's billion-dollar subsidies for industrial polluters.
- As of 2025, only electric cars or other vehicles with CO_2-free engines should be licensed. Such a measure works, as California showed in the '90s, by introducing quotas on electric vehicles. China, the world's biggest automotive market, is about to introduce such quotas from 2019 on. Now all the others must follow.
- Public transport must be expanded significantly. More conferences should take place via Skype instead of face-to-face meetings. Houses, streets and industry should occupy less space; we need to densify our cities upward and intelligently. Building ecologically does not mean new buildings, but primarily redeveloping and renovating. New industrial plants should be CO_2 emission-free from 2025 on. A deadline for compulsory intro-

duction of zero-emission technologies will drive the necessary innovations worldwide.

- About 25 percent of annual greenhouse gas emissions are caused by the production of food—especially meat products. Do you bear in mind that the production of beef soup causes ten times more greenhouse gases than vegetable soup does? Does meat soup really taste ten times as good as vegetable soup? So everybody should consider the guidelines of the German Nutrition Society (Deutsche Gesellschaft für Ernährung, DGE). The Society suggests reducing meat consumption first by half, later by two thirds. This helps to prevent obesity and high blood pressure, slows down climate change and reduces nitrogen levels in ground water. Climate change must also be understood as a medical emergency. The correlation between climate change and our health has so far received far too little attention. Global warming is the greatest threat to our health in the twenty-first century, according to the World Medical Association. And fine dust increases the risk of a stroke, asthma and diabetes. Global warming is fatal. Or vice versa: climate protection improves our health. By riding a bicycle or by walking, we not only take good care of the environment but also reduce the risk of cardiovascular diseases, diabetes and being overweight. Burning less coal means less fine dust and fewer lung patients. Unlike other religions, Buddhism does not grant us humans a higher right to life than other living beings. A Buddhist monk would never say, as in the Middle Ages the Christian monk Saint Thomas Aquinas did, "Animals have no soul." Neither would Jesus have used these words. He advocated compassion for all creatures. In the New Testament I found sixteen animal species in his parables.

- We must reforest worldwide and green the deserts, as the child and youth organization "Plant-for-the-Planet" has been doing for years in an exemplary manner. They have already planted over thirteen billion trees, with the aim of 1,000 billion trees. Young trees will not help immediately. But at least something is growing. Pakistan has announced that it will plant ten billion trees by 2030. Why does the German Ministry of Agriculture envisage planting only 100 million trees? Ethiopia, even poorer than Pakistan, holds the world record for planting trees: in the summer of 2019 over 350 million young trees on a single day!

- We must vote exclusively for politicians who honestly represent our interests rather than those of obsolete fossil-nuclear energy or of the fossil automobile industry. Democracy instead of autocracy and sun instead of atom and coal.

- Solar development in poor countries is the best precaution against population growth that is out of control.

- We all can make fewer purchases and reduce waste, ride a bike more often or go jogging, be more environmentally friendly whenever we have a party, switch to green electricity, invest our money green and fair. We should at last do what we deem right. Live in a simpler way so that others will simply survive. Think more and offer resistance against ignorance and shortsightedness. We can free ourselves from overabundance.

X.

What Can I Do?

1. Choose wisely

It takes great effort to realize these Commandments. But in the end it will mean a better life for all of us, a life worth living for. The fruit of climate justice is peace. In order to overcome a materialistic worldview, we need a positive vision that is more attractive than our obsolete worldview. The world revolution for compassion, as suggested here by the Dalai Lama, can be of great, perhaps even decisive, help. Now the time seems ripe. Large parts of the younger generation, as well as increasingly the older ones, seem prepared.

Who says that an individual can't do anything about it? If each and every one puts his or her own house in order, the whole world will be a better place. "The future depends on what you do today" (Mahatma Gandhi). Who, apart from ourselves, could stop us? A better world begins with each individual.

We are in danger of disappearance, unless we learn that a healthy forest ensures our own health. Perhaps we humans need to develop a greater awareness of trees. Novelist Richard Powers says we should immediately abandon our blindness toward the supposed "special status as a human being." An "awareness of plants" could be helpful. These terms are pretty close to the Dalai Lama's Buddhist holistic thinking and his message "Revolution of Compassion" as well as Albert Schweitzer's ethics: "Reverence for All Life." Whoever remains alert cannot deny the threatening catastrophe any longer. How far are we prepared to go in order to stop the apocalypse? Whoever is responsible for a problem can also put things right—probably the most important thesis of this book. As a human family we share a common destiny. So let's take care that it's not getting too hot.

We decide in favor of, or against, sustainable building, we decide whether we travel "green" or in a way that is damaging to the climate, whether we live on food that destroys resources or the produce of organic farming and whether we use renewable or fossil-atomic energy. The transformation we need for a good future for all of us has already taken place among millions serving as models. This transformation is no longer a dream; it is often becoming real.

Rolf Disch, solar architect, has been building houses for twenty-five years that produce three times as much solar electricity as the occupants need, as well as enough energy for the heating. His house earns money. This is the solar future. His own house produces more than six times as much solar electricity as he and his wife consume. I know social housing projects that are powered by solar energy. The occupants now spend half as much money on energy as they did in the past, when traditional energy was supplied. Solar electricity is social electricity.

This book is meant to encourage action. Political as well as private, personal action, including voting habits in favor of environment and climate, that is the basis of our lives. Our voting behavior reflects our political responsibility in a democracy. At the beginning of this book, the Dalai Lama says that in Europe he would vote for the Green Party.

I am also convinced that, without the Green Party participating in the Federal Government, Germany will clearly miss its climate protection targets in 2030, as in the case of 2020.

I say so as somebody who was a member of the Christian Democrats for twenty-eight years. The traditional parties are still stuck in structures and thinking of the old energy and automobile companies.

All technical requirements needed to overcome the fossil-atomic age are met. Our problem is not a lack of knowledge but of rapid implementation. With its technological advances in renewable energy, Germany has no reason to fear the forthcoming changes implied by the 100 percent solar energy transition. On the contrary: today's resources crisis offers us a unique chance. Alternative energy technology.

It is now exclusively a breakthrough, ensuring that our children and grandchildren can look to the future with self-confidence, optimism and joy. One day we should be able to say: children, this is your world; we have helped to make life beautiful, for you as well. Life is waiting for you.

We adults must now learn from children. We might finally grow up. It is high time everyone who thinks he is a grown-up oriented himself by the realism of the "Fridays for Future" movement. This means that each individual assumes responsibility, goes on strike, becomes politically active, learns to take himself seriously, and stops being childish, and adults finally become grown-ups. No more illusions of eternal growth. Growth for growth's sake is an idle objective. Ecosocial market economy strives for sustainability and quality of life for everybody. Questions like the following arise: What is more important, healthy and sufficient nutrition for all or even more cars, additional mobile phones and intercontinental air travel? It is time to face realities.

2. Is there still hope for us?

The Dalai Lama suggests that climate policy follows science, just as the "Fridays for Future" movement does. Young people have nothing to lose but their future. And please tell me what is more important than the future of our children and grandchildren?

Is there still hope for us? Yes, there is. But not for long!

Our transformation is not a blind fate. The future is what we make of it today. The most effective way to predict the future is to shape it. Looking back on my life, I would say that the ecosocial market economy is the most effective system in which billions of people can realize their dreams of a better world. The Paris Agreement of 2015 as well as the UN Millennium Development Goals are the basis of this new global ecosocial market economy. As a realist and journalist, I know of course that there is a big difference between writing about wonderful targets and their realization. Real evolution means change, modification, transformation, turnaround and future. There is much to do. Future means working for the future. And all this must be beautiful, aesthetic, attractive and not deterrent.

The last time a general strike took place in Germany was on November 12, 1948. "It's in your life's interest. Join in" was the motto on the posters of the trade unions. This general strike triggered the social market economy in Germany, which between the years 1950 and 1980 promised "prosperity for all," and this promise was kept in a remarkable way. This historic strike was the basis of the German economic miracle. Shorty afterward, Ludwig Erhard and his administration re-enforced social insurance schemes and introduced price regulations.

The global climate strike on September 20, 2019, could be the starting signal for a worldwide social-ecological market economy. In 163 countries, more than four million people took to the streets. They not only wrote history. Perhaps even more important, on their posters was written, "We will come back."

This first worldwide strike in human history was about nothing less than saving the world. On the same day the German government decided to apply climate protection measures—albeit only a minimum—and three days later the UN met to discuss climate protection. Never before had there been such a weekend: the world is rising up. The issue is finally where it belongs: in the center of international politics. Millions of young people are taking to the streets and the rulers of the world are under pressure: in Australia and India, in Germany and France, in the USA and in Bolivia, in Kenya, Bangladesh and in South Africa.

After World War II, fathers and mothers of the

social market economy in Germany were courageous and forward-looking. The well-known phrase "I think, therefore I am" of French Enlightenment thinker René Descartes should today be completed by Buddhist enlightened insight: "I live compassion, therefore I am," or by Carl Gustav Jung's message: "I dream, therefore I am." Most Indians, Chinese, Africans or South Americans will never understand "I think, therefore I am"—which is such an important phrase to us Europeans—nor our European atheism. Most people do not define themselves through reason, but through emotion, not through "Me" but through their relationship to fellow beings, i.e., through "Us."

Buddha and Jesus were the most important, most lasting and most convincing models to learn from in the past two and a half millennia. However, we have not learned enough from them. Otherwise, we would neither upgrade nuclear weapons nor wage wars, nor destroy the environment. Who finally prevents us from learning their teachings if not ourselves? What we can learn in ecological Buddha's teaching, as well as in ecological Jesus's teaching, is trust in the Creation. Jesus's decisive question for each of us: Do you trust in money or in God? In greed or love? His basic program is the "Sermon on the Mount." And Buddha's basic program is compassion with all life, which he teaches us in the "Eightfold Path."

Today's prevailing, near-global and unbridled materialistic neo-liberalism has in reality become a dictatorship of the international financial capital. The Dalai Lama comments: money is an important means of exchange, but "it is wrong to consider money a god or a substance endowed with some power of its own." And Peter Spiegel wrote the book, *WeQ—More than IQ. A Farewell to the I-Culture.* The "We-economy" shows new ways toward a more humane economy based on greater mindfulness, empathy, nonviolence, truthfulness, transparency and responsibility. These six virtues are the basis of the ethics of compassion, as advocated by Buddha and the Dalai Lama. These Buddhist economy ethics are similar to what Hans Küng, a Catholic theologian, calls "World Ethos" or the Protestant theologian Albert Schweitzer, "Reverence for Life." Fortunately there are already an increasing number of groundbreaking WeQ trends and projects, which I am going to outline in a separate book published by Benevento, in cooperation with Peter Spiegel. We will point out that this revolution of compassion is nothing but active altruism. That is to say: the gain of others pleases me as much as my personal gain. This opens our hearts. Such altruism is spiritually motivated. Supporting the well-being of all living creatures naturally includes my personal well-being. An economic theory that ingenuously identifies happiness with material wealth is at best naive.

As a mendicant monk, the Dalai Lama has practically no personal belongings. Nevertheless, he is regarded by millions as their happiest contemporary. Every day this religious leader meditates up to four hours. Training of the mind. That is: peace of mind, happiness of recognition and understanding of our illusions. "Inner tranquility is actually the source of happiness," according to economics professor Karl-Heinz Brodbeck.

September 20, 2019, can become a turning point in human history, at least the beginning of a turnaround. Climate protection, climate justice and solidarity take on a new meaning.

Is it all illusion? Who, in 2018, would have dared to predict that a Swedish teenager was about to bring new inspiration to the agenda of world politics?

★ ★ ★ ★ ★

Acknowledgments

In writing this book, the Office of His Holiness the Dalai Lama in Dharamsala, India, and the Gaden Phodrang Foundation of the Dalai Lama in Switzerland have been of invaluable help. I am very grateful that, for over thirty-five years, I had the chance to inform audiences about environmental issues on my radio and TV station, SWR, but also on Arte, 3sat and the Third Programmes. This information often had to be pushed through against the will of the ARD authorities. Today more colleagues than before keep raising these survival issues, for instance, Volker Angres, Harald Lesch or Sven Plöger.

Photo: Bigi Alt, who brought the two men together for the first time in 1982, the Dalai Lama and Franz Alt. In 1981 Bigi Alt shot a TV movie in Tibet secretly and without Chinese watchdogs. This film was broadcast by Franz Alt several times on German (ARD) and foreign TV stations. And this was the beginning of a lifelong friendship between the Dalai Lama and the Alt family.

For more information:
www.dalailama.com and www.sonnenseite.com.